CLIMATE CHANGE AND ITS CAUSES, EFFECTS AND PREDICTION

INTERNATIONAL FINANCING EFFORTS TO ADDRESS CLIMATE CHANGE

CLIMATE CHANGE AND ITS CAUSES, EFFECTS AND PREDICTION

Additional books in this series can be found on Nova's website under the Series tab.

Additional e-books in this series can be found on Nova's website under the e-books tab.

Climate Change and its Causes, Effects and Prediction

International Financing Efforts to Address Climate Change

Randy N. Matthews
Editor

Nova Science Publishers, Inc.
New York

For permission to use material from this book please contact us:
Telephone 631-231-7269; Fax 631-231-8175
Web Site: http://www.novapublishers.com

NOTICE TO THE READER

Library of Congress Cataloging-in-Publication Data

International financing efforts to address climate change / editor, Randy N. Matthews.
p. cm.
Includes index.
ISBN 978-1-62100-702-9 (hbk.)
1. Climatic changes--Economic aspects. 2. Climatic changes--Government policy. 3. Environmental policy--International cooperation. I. Matthews, Randy N.
QC903.I4925 2011
363.738'74--dc23

2011037418

Published by Nova Science Publishers, Inc. † New York

CONTENTS

PREFACE

Many voices, domestic and international, have called upon the United States to increase foreign assistance to address climate change. Proponents maintain that such assistance could help promote low-emissions and high-growth economic development in lower income countries, while simultaneously protecting the more vulnerable countries from the effects of a changing climate. Recent studies estimate the needs for climate change financing in the developing world to range from US$4 billion to several hundred billion annually by the year 2030. This book examines the needs, sources and delivery methods for international climate change financing with a focus on the fundamental dispute concerning who should pay for it and how.

Chapter 1- Many voices, domestic and international, have called upon the United States to increase foreign assistance to address climate change. Proponents maintain that such assistance could help promote low-emissions and high-growth economic development in lower income countries, while simultaneously protecting the more vulnerable countries from the effects of a changing climate. Recent studies estimate the needs for climate change financing in the developing world to range from US $4 billion to several hundred billion annually by the year 2030. The United States has pledged funds in such fora as the United Nations Framework Convention on Climate Change (UNFCCC, 1992), the Copenhagen Accord (2009), and the UNFCCC Cancun Agreements (2010), wherein the wealthiest countries, in aggregate, agreed to provide up to $30 billion in "fast start" financing for the 2010-2012 period and to mobilize $100 billion annually by 2020. Pledged funds are to come from a wide variety of sources, both public and private, bilateral and multilateral, including alternative sources of finance. Lower income countries have sought assistance that is *new, additional to previous flows, adequate, predictable,* and *sustained.*

Chapter 2- The United States supports international financial assistance for global climate change initiatives in developing countries. Under the Obama

Administration, this assistance has been articulated primarily as the Global Climate Change Initiative (GCCI), a major strategic platform within the President's 2010 Policy Directive on Global Development. The GCCI aims to integrate climate change considerations into relevant foreign assistance through a full range of bilateral, multilateral, and private sector mechanisms to foster low-carbon growth, promote sustainable and resilient societies, and reduce emissions from deforestation and land degradation. The GCCI is implemented through programs at the Department of State, the Department of the Treasury, and the U.S. Agency for International Development (USAID), and is funded through the Administration's Executive Budget, Function 150 account, for State, Foreign Operations, and Related Programs.

Chapter 3- The United States contributes funding to various international financial institutions to assist developing countries address global climate change and other environmental concerns. Congress is responsible for several activities in this regard, including (1) authorizing periodic appropriations for U.S. financial contributions to the institutions, and (2) overseeing U.S. involvement in the programs. Issues of congressional interest include the overall development assistance strategy of the United States, U.S. leadership in global environmental and economic affairs, and U.S. commercial interests in trade and investment. This report provides an overview of two of the larger and more recently instituted international financial institutions for the environment—the Climate Investment Funds (CIF)—and analyzes their structure, funding, and objectives in light of the many challenges within the contemporary landscape of global environmental finance.

Chapter 4- Over the past several decades, the United States has delivered financial and technical assistance for climate change activities in the developing world through a variety of bilateral and multilateral programs. The United States and other industrialized countries committed to such assistance through the United Nations Framework Convention on Climate Change (UNFCCC, Treaty Number: 102-38, 1992), the Copenhagen Accord (2009), and the UNFCCC Cancun Agreements (2010), wherein the higher-income countries pledged jointly up to $30 billion of "fast start" climate financing for lower-income countries for the period 2010 – 2012, and a goal of mobilizing jointly $100 billion annually by 2020. The Cancun Agreements also proposed that the pledged funds are to be new, additional to previous flows, adequate, predictable, and sustained, and are to come from a wide variety of sources, both public and private, bilateral and multilateral, including alternative sources of finance.

In: International Financing Efforts ...
Editor: Randy N. Matthews
ISBN: 978-1-62100-702-9

Chapter 1

INTERNATIONAL CLIMATE CHANGE FINANCING: NEEDS, SOURCES, AND DELIVERY METHODS

Richard K. Lattanzio and Jane A. Leggett

ABSTRACT

Many voices, domestic and international, have called upon the United States to increase foreign assistance to address climate change. Proponents maintain that such assistance could help promote low-emissions and high-growth economic development in lower income countries, while simultaneously protecting the more vulnerable countries from the effects of a changing climate. Recent studies estimate the needs for climate change financing in the developing world to range from US $4 billion to several hundred billion annually by the year 2030. The United States has pledged funds in such fora as the United Nations Framework Convention on Climate Change (UNFCCC, 1992), the Copenhagen Accord (2009), and the UNFCCC Cancun Agreements (2010), wherein the wealthiest countries, in aggregate, agreed to provide up to $30 billion in "fast start" financing for the 2010-2012 period and to mobilize $100 billion annually by 2020. Pledged funds are to come from a wide variety of sources, both public and private, bilateral and multilateral, including alternative sources of finance. Lower income countries have sought assistance that is *new, additional to previous flows, adequate, predictable,* and *sustained.*

The fundamental dispute concerning international financing for climate change centers upon who should pay for it and how. The debate has been dominated by economic assessments of market- based mechanisms aimed at changing price incentives so that investment in low-emissions development becomes more attractive (e.g., cap and trade, carbon fees, loan guarantees). Many agree that private sector investment will likely have a significant role to play in any low- emissions future, and that establishing a price on GHG emissions will likely have a part in any effective policy agenda. However, concerns remain whether such mechanisms can induce the required shifts in production and consumption patterns, mobilize the necessary investment, and contribute adequately to international financial assistance. From this perspective, public funds—including from national governments and international organizations—continue to be a key driver for climate change investment, specifically in low income countries.

Many methods for disbursing international climate change financing currently exist. All have a role in catalyzing climate action. They include private sector funding through such avenues as foreign direct investment (FDI), export credit markets, multilateral development banks and finance corporations, and the various U.N. Kyoto Protocol market mechanisms, as well as public sector funding through official development assistance (ODA), multilateral trust funds (e.g., the Global Environment Facility (GEF), Climate Investment Funds (CIF), Green Climate Fund (GCF)), and the concessional lending windows housed at the World Bank Group. Many contend that the financial architecture is underfunded, unnecessarily complex, and lacks both strategic mandate and adequate coordination. Debate has arisen over the proper financial instruments to employ in lower income countries as well as the role shared by the public and private spheres.

Up to this point, the United States has relied mostly on direct budget appropriations to finance climate change actions internationally, but recent Congresses have considered several alternatives that could generate new financing for international purposes. Many in Congress and the public at large may question why the United States should help finance other countries' efforts on climate change. Some claim that international financing would incur costs to the United States, or redirect funds that could be used for domestic purposes and send them overseas. Others, however, contend that international financing may offer potential benefits to the United States in terms of global environmental protection, expanded commercial markets, and increased national security.

INTRODUCTION

While some investigators disagree with the current risk assessments of climate science, most accept the findings of the U.S. National Academies that the Earth's climate has changed over the past century and that human activities—particularly emissions of greenhouse gases (GHG) through such activities as fossil fuel use, agricultural practices, and deforestation—have very likely caused most of the observed effects.[1] Broadly agreed findings conclude that the world faces risks from the damaging effects of a changing climate unless GHG emissions are limited. There is some political consensus internationally to try to stabilize GHG concentrations in the atmosphere at approximately 450 parts per million by volume (ppm)[2] carbon dioxide-equivalent (CO_2e),[3] a volume which is projected to limit global warming to around 2°C above pre-industrial levels. Some people argue that a 2°C target, if attained within a sufficient time frame, might prevent dangerous anthropogenic interference with the climate system, allow ecosystems to adapt naturally to climate change, ensure that food production is not threatened, and enable economic development to proceed in a sustainable manner.[4] Others suggest different target levels of varying stringency.

Large-scale financial investments are projected to be needed to meet the global demand for energy, water, transportation, heating, and other infrastructure services in countries with growing populations and rising incomes. Financial requirements would be increased if countries made such investments with the additional consideration of addressing climate change. Proponents maintain that climate-relevant investments would promote low-emissions, high-growth economic development while simultaneously protecting the more vulnerable countries and communities from the effects of climate change. Some countries have begun to make moderate adjustments focusing, in particular, on energy efficiency strategies, low-emissions energy infrastructure, and sustainable land use, land use change, and forestry practices. However, despite expressions of concern and commitment, the shift toward climate-relevant investment has been deemed slow by many. Further, while industrialized countries continue to contribute a disproportionally large share of global GHG emissions, and historically have contributed most of the global GHG emissions over the past two centuries, future emission growth is projected to arise mostly from the developing world, whose populations and economic aspirations continue to grow. Thus, even if developed country emissions are significantly curtailed, climate targets may not be met without the deployment of similar abatement efforts in the developing world.[5]

While there is little doubt that the most efficient GHG reduction strategy would require efforts from all countries, a fundamental dispute centers upon who should pay for them and how. Most, if not all, low income countries have argued that their success in abating GHG emissions and curtailing deforestation would depend critically on receipt of international financial and other support. They argue that reducing their share of GHG emissions and adapting to the effects of climate change would incur costs above and beyond their normal economic growth trajectories. These costs are particularly challenging to countries that have low incomes compared to industrialized nations, consider alleviating poverty as their first priority, and conclude that they have contributed only a minor share of the historical GHG emissions that force climate change.

Higher income countries, including the United States, have pledged financial assistance to lower income countries in such fora as the United Nations Framework Convention on Climate Change (UNFCCC, 1992).[6] While vaguely defined, the UNFCCC pledges are not voluntary commitments, but treaty obligations, signed and ratified by the U.S. government.[7] More recently, the Copenhagen Accord (2009) stipulated—and the UNFCCC Cancun Agreements (2010) restated—that the wealthiest countries[8] in aggregate would commit to provide up to $30 billion "fast start" financing in the 2010-2012 period and to mobilize $100 billion annually by 2020 to promote mitigation, adaptation, technology transfer, and capacity building efforts in lower income countries. The funding is to come from "a wide variety of sources, [both] public and private, bilateral and multilateral, including alternative sources of finance."[9] The Copenhagen Accord is a non-binding political agreement among countries; the implementation of the Cancun Agreements, as a product of the UNFCCC Ad Hoc Working Group on Long-term Cooperative Action under the Convention (AWG-LCA), is still under negotiation by Parties.

This report aims to inform Congressional decision-making on U.S. provisions for financial assistance to low income countries to address climate change. It proceeds by first framing some perspectives on the U.S. role in international financial assistance for climate change. It then addresses the following questions:

1. How much funding might be needed to address the problem?
2. What might the funding be used for?
3. Where might the funding come from?
4. How might the funding be delivered?

The final section of the report summarizes past and current U.S. contributions to international climate change initiatives.

PERSPECTIVES ON THE U.S. ROLE IN INTERNATIONAL CLIMATE CHANGE FINANCIAL ASSISTANCE

Calls continue domestically and internationally for high income countries to increase financial assistance to low income countries to address climate change. Many in Congress and the public at large may question why the United States should help finance other countries' efforts to reduce GHG emissions or to adapt to climate variability and change. Some claim that international financing would incur costs to the United States, or redirect funds that could be used for domestic purposes and send them overseas. Others, however, contend that international financing may offer potential benefits to the United States in terms of environmental protection, expanded commercial markets, and national security.

Below is a brief outline of some of the arguments in support of and in opposition to the role of the U.S. government in foreign aid in general and international climate change assistance in particular.

In Opposition

Fiscal Constraints: Some critics of international climate change assistance to lower income countries argue that the United States needs to retain available funds for domestic priorities, such as fostering renewed economic growth and creating jobs. They contend that the United States should not be burdened with higher taxes or prices for investments abroad.[10] The burden is exacerbated during times of economic downturns, when governments are hard-pressed to generate fiscal resources to adequately address domestic challenges and maintain basic levels of public services and quality of life. For those who support some form of international development assistance, they may argue that aid should be directed to country priorities other than climate change, such as improving public health systems, water and sanitation resources, or infrastructure.

Misuse of Funds: Many critics have claimed that international financial assistance targeted to help average citizens in low income countries often ends up supporting inefficient and bloated bureaucracies (e.g., recipient governments, donor-funded multilaterals, or non-governmental organizations). They argue that the national and international institutions that dispense financial assistance focus on "getting money out the door" to lower income countries, rather than on delivering services; emphasize short-term outputs like reports and frameworks but do not engage in long-term activities like the evaluation of projects after they are completed; and put enormous administrative demands on lower income country governments.[11] Bilateral and multilateral development agencies have also been criticized for the fragmentation of foreign assistance across many small and uncoordinated bureaucracies, the lack of transparency about project procurement practices and operating costs, and the proportion of funds that is misused or lost through instances of graft, corruption, and other political inefficiencies.[12]

Poor Results: Some critics contest the overall effectiveness of foreign assistance in spurring economic development and reform in low income countries. Many studies have examined the effects of international assistance provided to lower income countries, including both bilateral and multilateral mechanisms, and have returned mixed results. Conclusions range from ineffective,[13] to highly effective,[14] to effective "in some countries under specific circumstances."[15] The divergent results of these studies may make it difficult to reach firm conclusions and support continued contributions. Further, some commentators find that international environmental assistance poses even greater uncertainties than other forms of aid due to difficulties in assessing, measuring, reporting, and verifying environmental indicators. Some statistics have confirmed these perceptions, as the success rate for environmental projects is often far below those of education, health, or infrastructure.[16]

Development Inefficiencies: There is no satisfactory metric on the effectiveness of international financial assistance to lower income countries. Some critics claim that it may do more harm than good.[17] They assert that a reliance on foreign capital fosters an unhealthy economic dependence, making poor countries poorer and economic growth slower, and leaving recipients more debt-laden, inflation-prone, vulnerable to the vagaries of the currency markets, and unattractive to higher-quality investment. They claim that grant-based assistance as a platform for development is contradictory to sustainable economic growth and can squash private sector efforts in commercial markets. They claim debt-based assistance is worse, requiring loans to be repaid at the expense of recipient country's education, health, and infrastructure investments. Further, large inflows of foreign capital may have the effect of killing off a country's export sector by causing domestic currency to strengthen against foreign ones (referred to by economists as "Dutch Disease"). As a counter to the practice of international financial assistance, these commentators promote a strategy of development that emphasizes the role of entrepreneurship and private markets over an aid system based on cycles of transfer flows.

Lack of Consensus on Climate Science: Some critics point to scientific uncertainties and ambiguities within the fields of atmospheric chemistry and climatology as reasons to postpone and/or reconsider international climate change assistance policies and programs. They contend that the current scientific findings on climate change may not be sufficient to warrant government action, either domestic measures to mitigate GHG emissions and adapt to the effects of climate change or international policy actions and financial assistance to support other countries' efforts.

In Support

Commercial Interests: Some advocates argue that international climate change assistance to lower income countries to support low-emission economic growth could benefit U.S. businesses through increased trade, commerce, and economic activity in the global marketplace. They contend that American clean energy and environmental management companies are well positioned to provide the innovative technology and services needed to meet the rapidly growing demand in emerging economies. Increased financing would not only promote development in the host country, but allow U.S. industries to make competitive inroads into rapidly expanding markets, improve the advancement and commercialization of U.S. technologies, mobilize greater investment in domestic sectors, and enhance job creation in the United States. Decreased funding may cede American influence in global markets to other economic powers still engaged with lower income countries on environmental and natural resource issues (e.g., the European Union, China).

Investment Efficiencies: Some advocates claim that the costs of responding to tomorrow's climate-related catastrophes, instabilities, conflicts, and technological needs would be much higher than the costs of working today to prevent them through emissions reductions. "Each year of delay will lock in an increased amount of old [i.e., high-emissions] technology," according to the United Kingdom's Secretary of State for Energy and Climate Change.[18] Economists often note that lower income countries account for nearly all of the recent growth in global emissions and represent the cheapest opportunity to mitigate GHG pollution as part of a cost-effective solution. Additionally, some are concerned that locking in the developing world to a reliance on older, more GHG-intensive technologies during economic development may limit their flexibility and increase their costs to respond efficiently to future pollution abatement or climate resilient strategies. (Sometimes this is called a problem of "stranded capital.") Economists view some level of early investment as efficient, to hedge against these future risks.[19]

Natural Disaster Preparedness: Some advocates point to international climate change assistance as a means to assist in global disaster preparedness. They claim that recorded natural disasters continue to increase each year resulting in more casualties and mounting economic losses.[20] They assert that extreme weather events have lead to increased droughts, food shortages, and resource competition, which, in turn, has lead in some cases to population displacement and migration. Some have proposed that increased international assistance to climate change adaptation programs could help avoid capital and other losses (e.g., buildings, infrastructure, etc.), minimize the redirection of strategic resources to ad hoc disaster response and urgent humanitarian needs, and avoid chronic humanitarian crises, such as food shortages, particularly for the resource poor in the least developed countries.[21]

National Security: Some advocates argue that international climate change assistance could help address risks to national security. According to a 2008 National Intelligence Assessment, the impacts of global climate change may worsen problems of poverty, social tensions, environmental degradation, and weak political institutions across the developing world.[22] In October 2010, Chairman Mullen of the Joint Chiefs of Staff highlighted that climate change creates conditions "that could lead to failed states and make populations vulnerable to radicalization."[23] Combating environmental drivers such as climate change, desertification, biodiversity loss, and deforestation could reduce the instability caused by the scarcity of, and potential competition for, resources like water, food and habitat. Some see international financial assistance for climate change as a means to help make lower income countries less susceptible to these threats, for the benefit of both the lower income country and the security interests of the United States.

International Leadership: Some advocates contend that international climate change assistance to lower income countries helps the United States improve its leadership in global environmental issues. Through leadership, the United States may be able to influence and set important international economic and environmental policies, practices, and standards. But leadership— apropos of voting share—is tied directly to the level of financial contribution in many multilateral organizations. Some believe that delivering on financial pledges is an important opportunity for the United States to demonstrate credibility and support negotiations not just on environment, but on economic and security issues as well. Withdrawing commitments may cede American influence in world affairs to other economic powers still engaged with lower income countries on environmental and natural resource issues (e.g., the European Union, China). Weakened influence could manifest in challenges to political negotiations, economic relationships, trade preferences, and future collaborations.

International Obligations: Many advocates stress that the commitment to international climate change assistance to lower income countries is codified in current multilateral agreements. Under the United Nations Framework Convention on Climate Change (UNFCCC), signed in 1992, the United States and other industrialized countries listed in Annex II of the Convention[24] committed to provide financial and technical assistance to help lower income countries' efforts to meet their UNFCCC obligations. While these commitments are legally binding, they are vaguely defined, making them impractical to quantify and enforce. More recent negotiations have striven to produce more quantified figures, and the Copenhagen Accord of 2009 produced an agreement by the wealthiest countries to provide $30 billion of "fast start" financing in the period 2010 to 2012 and to seek $100 billion annually by 2020.[25] However, these financial targets are not legally binding, and accounting and enforcement remain difficult.

Equity Issues: Some advocates consider international climate change assistance to lower income countries a moral responsibility and a matter of climate equity. Not only have today's high income economies generated about 80% of past fossil fuel-based emissions, but those same emissions have helped carry them to high levels of social and economic well-being. Past behavior arguably calls for the industrialized countries to provide funding to reduce the current and future risks imposed on others.

Table 1. Various Estimates of Annual Climate Change Financing Needed for Mitigation and Adaptation Activities in Lower Income Countries in Constant 2008 US$ Billions

Sources of Estimates	Estimate 2010-20	Estimate 2030	Details
Mitigation Costs			
Project Catalyst (2009)	45	—	See Figure 4 below
McKinsey & Co. (2009)	—	190	Total costs if pursuing the most economically rational abatement opportunities to their full potential with transaction and program costs
Pacific Northwest National Laboratory (PNNL) (2008)	—	151	—
Mitigation Financing			
Project Catalyst (2009)	70-100	—	See Figure 4 below
McKinsey & Co. (2009)	326	612	Total upfront investment needed incre-mental to business-as-usual (BAU)
International Institute for Applied System Analysis (IIASA) (2009)	68-179	287	—
International Energy Agency (IEA), Energy Technology Perspectives (2008)	—	614	Annual averages through 2050
Adaptation Costs			
World Bank (2009)	10-45	—	Cost of climate-proofing development assistance, foreign and direct investment
United Nations Development Program (UNDP) (2008)	90-114	—	Cost of climate-proofing development assistance, foreign and direct investment, plus cost of poverty reduction strategies and disaster relief programs
Stern Review (2006)	4-40	—	Cost of climate-proofing development assistance, foreign and direct investment

Sources of Estimates	Estimate 2010-20	Estimate 2030	Details
World Bank EACC (2009)	—	82-110	Average annual adaptation costs from 2010 to 2050 in the agriculture, forestry, fisheries, infrastructure, water resource management, and coastal zone sectors, including impacts on health, ecosystem services, and the effects of extreme weather events
Project Catalyst (2009)	—	16-40	2030 cost of capacity building, research, disaster management and the UNFCCC sectors (most vulnerable countries and public sector only)
UNFCCC (2008)	—	30-73	2030 cost in agriculture, forestry, water, health, coastal protection, and infrastructure

Sources: Project Catalyst, *Toward a Global Climate Agreement: Synthesis Briefing Paper,* June 2009; and World Bank, *World Development Report: 2010: Development and Climate Change,* 2009, with data for mitigation sourced from IIASA, 2009 and additional data provided by V. Krey; IEA, 2008; McKinsey & Company, 2009, and additional data provided by McKinsey (J. Dinkel) for 2030, using a dollar-to-Euro exchange rate of $1.25 to €1.00; PNNL figures from Edmonds and others, 2008, and additional data provided by J. Edmonds and L. Clarke; and for adaptation sourced from Agrawala and Fankhauser, *Economic Aspects of Adaptation to Climate Change: Costs, Benefits, and Policy Instruments*, OECD, 2008, except World Bank EACC, *Economics of Adaptation to Climate Change,* 2009.

Notes: Estimates are for stabilization of greenhouse gases at 450 ppm CO_2e, which would provide a 40%-50% chance of staying below 2°C warming by 2100. All figures have been adjusted to 2008 US$ in billions using U.S. Bureau of Economic Analysis (BEA) Gross Domestic Product (GDP) Deflator and a dollar-to-Euro exchange rate of $1.25 to €1.00.

COST ESTIMATES FOR INTERNATIONAL CLIMATE CHANGE ACTIVITIES

The financial costs of coping with climate change may reach trillions of dollars. These costs would aim to address some combination of mitigation activities (i.e., actions taken to eliminate or reduce the long-term risk and hazards of climate change) and adaptation activities (i.e., actions taken to adjust to climate change, moderate potential damage, or cope with the consequences). Estimates of the projected costs vary widely depending upon assumptions made about the accepted levels of pollution, the ambitiousness of the global response, its structure, timing, and implementation, the potential climate-related damages, the affected sectors, as well as the methods of sourcing and delivering the necessary funds. The stricter the emissions target, the higher the estimate. The longer the response is delayed, the more threatening the damages may be, and the greater the resources required to respond to the threats. Bearing the costs of action (or inaction) are individuals, firms, local communities, national governments, and/or the international community.

Findings

A variety of international institutions and non-governmental organizations have used various climate change mitigation and adaptation analyses to estimate the climate-related financing needs for lower income countries.[26] Table 1 reports findings from a variety of studies on both the "net costs" and "associated financing requirements" for mitigation and adaptation efforts in lower income countries. "Mitigation costs" refer to net incremental costs as factored over the lifecycle of the investment; and "mitigation financing" refers to the up-front capital investment needed over and above the business-as-usual (BAU) investment. Both mitigation costs and mitigation financing are estimated for stabilizing atmospheric concentrations at 450 ppm CO_2e; adaptation investments are estimated for a variety of assistance categories for both 2010 and 2030.

Focusing on the 450 ppm target, mitigation net costs in lower income countries range between \$150 billion and \$190 billion a year by 2030. If the estimates for associated financing needs are included, the total costs increase to \$287 billion to \$614 billion a year. For adaptation, the most comparable estimates are the medium-term figures produced by the UNFCCC and the World Bank, which range from \$30 billion to \$110 billion annually.

Currently, resources committed to address mitigation and adaptation in lower income countries cover approximately 5% of the aforementioned estimates. One recent study has contributions for mitigation-specific assistance at approximately \$20 billion annually. Very few studies have addressed current contributions for adaptation assistance.[27] Many claim that assessing climate- related financial and investment flows is a formidable challenge, given the inconsistencies across reporting systems, the many data gaps (with the further challenge of identifying the contributions of underlying finance, which unlike specific climate finance is not reported as such), and the complex web of flows (with the possibility of double counting).

Methodology

Estimates of the total net costs and financing requirements needed for mitigation and adaptation activities in lower income countries are often based on economic models of pollution abatement costs and/or adaptation investments aggregated across sectors and regions. This section discusses in more technical detail the economic modeling behind the various cost estimates presented in the previous section. Due to characteristic differences, estimates of net costs and financing requirements are often differentiated between the costs of *mitigation* and the costs of *adaptation.*

Mitigation Costs: Net Costs

Mitigation costs refer to the costs of actions taken to reduce or reverse the forces that contribute to global climate change. In higher and lower income countries alike, mitigation measures aim to reduce current levels of emissions and to emphasize low-GHG development. Strategies include transitioning to a low-emissions energy supply; capturing the opportunities in energy efficiency improvements in buildings, transportation, and industry; reducing deforestation and improving sustainable forest management to better serve as GHG emissions sinks; and employing more low- emissions and sustainable agriculture practices. In the future, it could also entail actions that remove carbon dioxide from the atmosphere and sequestering it permanently, or other geoengineering[28] technologies.

Estimates have been made of the incremental costs[29] of various mitigation strategies. Figure 1 shows one example of what analysts refer to as an "Emissions Abatement Cost Curve." The purpose of the curve is to summarize the many emission reduction strategies available, and to characterize their emission reduction potential as well as their net costs, if the strategy is "pursued aggressively." The curve charts the amount of potential reductions (expressed along the horizontal axis in gigaton[30] of CO_2e abated per year) against the cost of specific measures (expressed along the vertical axis in cost per ton of CO_2e abated).[31]

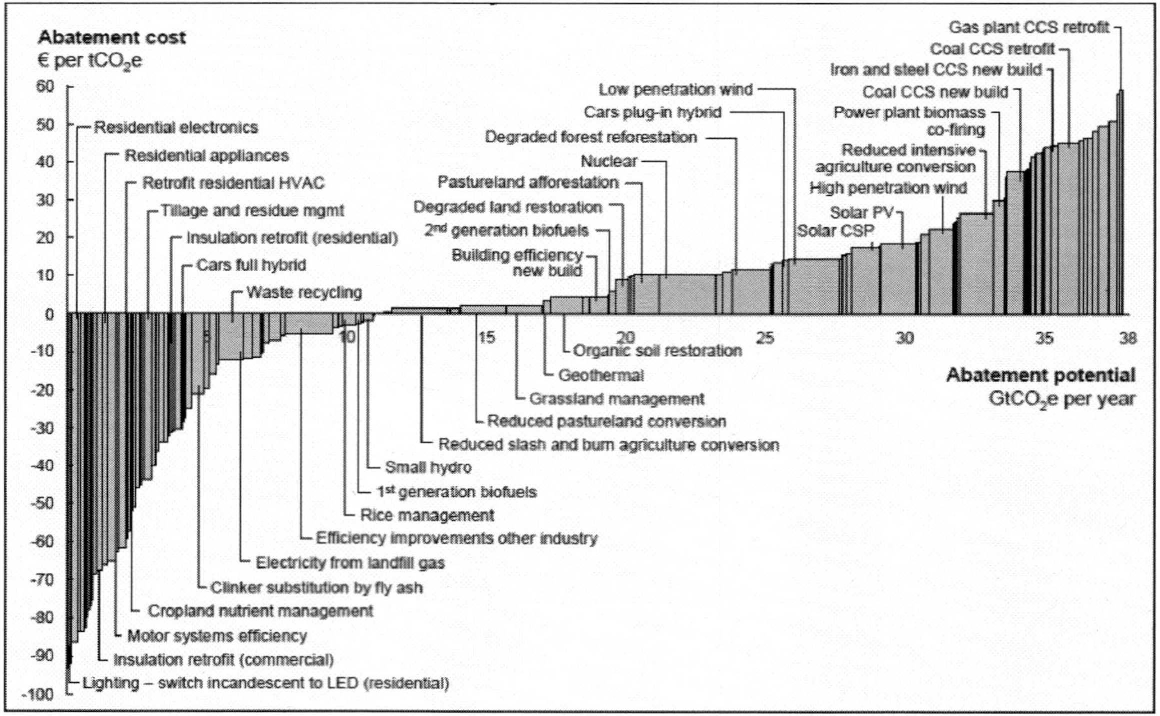

Source: McKinsey & Co, Pathways to a *Low-Carbon Economy*, v.2.0, 2009.

Notes: The curve presents an estimate of the maximum potential of all technical GHG abatement measures below €60 per tCO_2e if each lever was pursued aggressively. It is not a forecast of what role different abatement measures and technologies will play.

Figure 1. Global GHG Abatement Cost Curve Beyond Business-as-Usual for 2030.

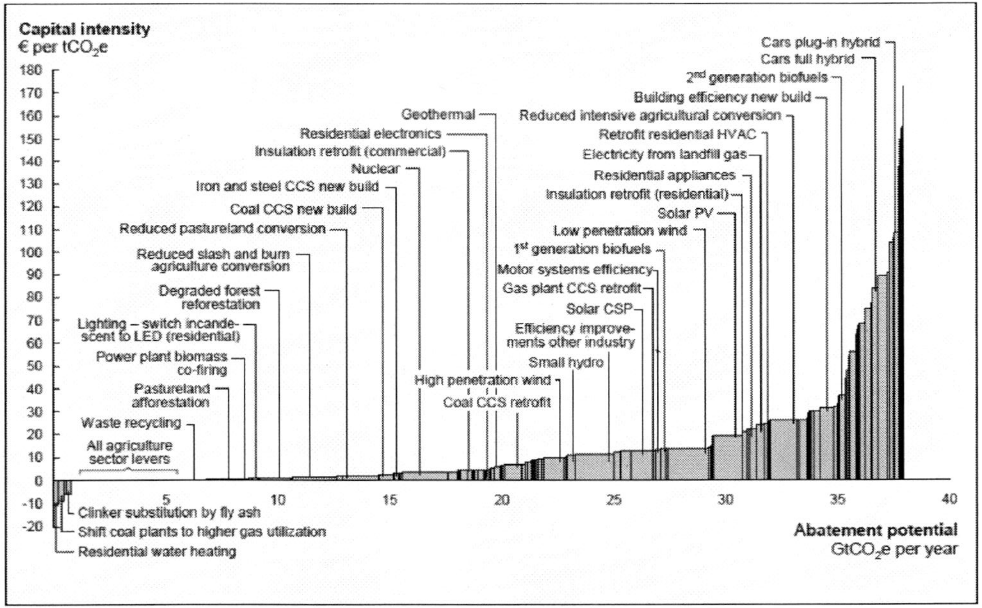

Source: McKinsey & Co, *Pathways to a Low-Carbon Economy*, v.2.0, 2009

Notes: The capital intensity of an abatement measure is defined as the additional upfront investment relative to the BAU technology, divided by the total amount of avoided emissions over the lifetime of the asset. For a more fuel efficient car, for instance, the capital intensity would be calculated as the additional upfront investment compared to the BAU technology, divided by the amount of CO_2 saved through lower fuel consumption during the lifetime of the car. The main difference with abatement cost is that the capital intensity calculation does not take financial savings through lower energy consumption into account.

Figure 2. Global Capital Intensity Cost Curve.

Of particular note, the curve shows that many abatement opportunities exist with net negative costs (approximately one-third of potential reductions), meaning that the measures may pay for themselves within the useful lifetime of the investment through efficiency savings (e.g., switching from incandescent light bulbs to LEDs would both reduce emissions and save money in the long run due to the extended product life and energy efficiency of LEDs). These negative cost options (shown on the left of the graph) are mainly found in energy efficiency measures in building, trans-portation, and industry as well as some fuel switching, recycling, and waste management practices. For emission reduction strategies in the agri-culture and forestry sectors (e.g., improved agriculture practices, afforestation, reforestation),[32] most options have low to moderate costs. Many of the low to moderate cost options fall within the curve's margin for error and are arguably breakeven estimates. Sectors with relatively high cost reduction opportunities are in some energy production options, with some emerging technologies having even higher costs than are represented on the graph given their nascent state of development.

Figure 1 demonstrates that ideally if all possible mitigation measures were taken on the cost curve in strict order from lowest-cost to higher-cost in sequence (i.e., from replacing all residential incandescents with LEDs all the way up to retrofitting all coal-fired power generation with carbon capture and storage technologies), global emissions abatement of 38 $GtCO_2e$ could be achieved in 2030.[33] At this rate, the average cost of the abatement opportunities would be \$5 per tCO_2e in 2030, and the total cost for realizing the whole curve would be \$187.5 billion in the year 2030. Transaction and program costs—not represented on the curve—are often estimated at an average of between \$1.25 and \$6.25 per tCO_2e abated, making the total annual global cost approximately \$250 billion to \$375 billion by 2030.[34] While an abatement cost of \$6 to \$12 per tCO_2e is reasonable in light of current economic discussions on climate change, many commentators stress that mitigation costs are extremely sensitive to policy choices. They increase steeply with the stringency of the emission reduction target and with the desired degree of certainty of reaching it. Global mitigation costs would likewise rise to the degree that the world deviates from a least-cost emission pathway (e.g., not tapping low cost reductions in lower income countries in the initial mitigation effort would increase global costs significantly). Further, the failure to allow for all mitigation opportunities would likewise increase overall costs (e.g., only concentrating on energy efficiency measures and not on forest and agricultural management could increase overall costs).

Mitigation Costs: Capital Intensity, or Up-Front Costs

Many commentators note that low-emission investments for mitigation activities often have high up-front capital costs, followed later by overall savings in operating costs.[35] Figure 2 charts the incremental capital needs, or "capital intensity costs" (i.e., the extra investment needed at the onset of a project over and above the business-as-usual (BAU) technologies), as an alternative metric to the net cost curve of Figure 1 (which charts the additional net costs of a low-emission project *over its entire lifetime*). Figure 2 shows that in some cases the difference between capital intensity costs and net costs can be as much as a factor of two to four, depending on the rate of opportunity costs assumed. The McKinsey & Company estimate concludes that the total global upfront investment needed for abatement measures would be $1,012.5 billion per year in 2030— incremental to BAU investments. For financially constrained countries, specifically many lower income countries, these high up-front capital costs can be a significant disincentive to invest in low-emission technologies.

The McKinsey & Company estimate outlined above is just one of many assessments conducted on financial needs. See Table 1 above for other mitigation cost estimates from a variety of studies.

Adaptation Costs

Adaptation costs refer to the costs of adjustments made in natural or human systems in response to actual or expected climate change and its effects. Estimates of adaptation costs have focused on the additional amount of investment needed to reduce the impact of anticipated future damages caused by climatic trends or events, including measures to increase resilience, reduce the impacts of anticipated disasters, and cope with the aftermaths. Examples of adaptation measures include employing climate-resistant crop varieties, improving irrigation systems, integrating sustainable land management into agricultural planning, protecting water resources, managing coastal zones, designing infrastructure for extreme weather or for sea-level rise, and improving public health services.

From a global perspective, the adaptation challenge may be greatest in the developing world. Lower income countries are generally more vulnerable to climate change because their economies are more dependent on climate-sensitive sectors such as agriculture, fishing, and tourism.

The figure plots a composite index of physical impact (taken as a function of climate sensitivity and climate-change exposure and derived from a number of global impact studies) against a composite index of social capacity (derived from a number of socioeconomic indicators).

Social capacity and vulnerability, as measured by projected impacts, are composite indexes of the indicators described in the table below.

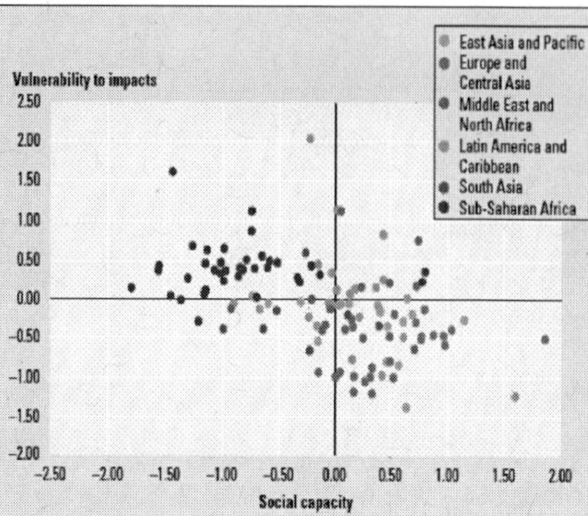

	Indicator	Metric	Source	Assumptions
Impact	Sea-level rise	Percent population affected by 1 meter rise	Dasgupta and others 2007	Landlocked countries assumed to experience zero impact
	Agriculture	Percent yield loss in 2050, IPCC SRES scenario A2b	Parry and others 2004	Decreasing yields represent decreasing welfare for country. Increased yields from climate change represent increasing welfare. Farm-level adaptation present
	Health	Percent additional deaths in 2050	Bosello, Roson, and Tol 2006	Additional deaths representative of all health impacts from climate change
	Disaster	Percent population killed by disasters (historical data set)	CRED 2008	Current disaster patterns to represent future areas at risk
Social capacity	Literacy	Percent population, aged >15 years, literate (1991–2005)	World Bank 2007c	The higher the literacy rate, the higher the social capacity
	Age dependency ratio	Ratio of dependent population to working population (2006)	World Bank 2007c	The lower the age dependency ratio, the higher the social capacity
	Primary completion rate (female)	Percent female population completing primary education (1991–2006)	World Bank 2007c	The higher the completion rate, the higher the social capacity
	Gini	Gini coefficient (latest available year)	World Bank 2007c	The lower the inequality, the higher the social capacity
	Domestic credit to private sector	Domestic credit to private sector, as percent of GDP (1998–2006)	World Bank 2007c	The greater the investment, the higher the social capacity
	Governance	WGI (World Governance Indicator) voice and accountability	Kaufman, Kraay, and Mastruzzi 2008	The higher the WGI score, the higher the social capacity

Source: World Bank, World Development Report 2010: Development and Climate Change, 2009.

Figure 3. Climate Vulnerability vs. Social Capacity.

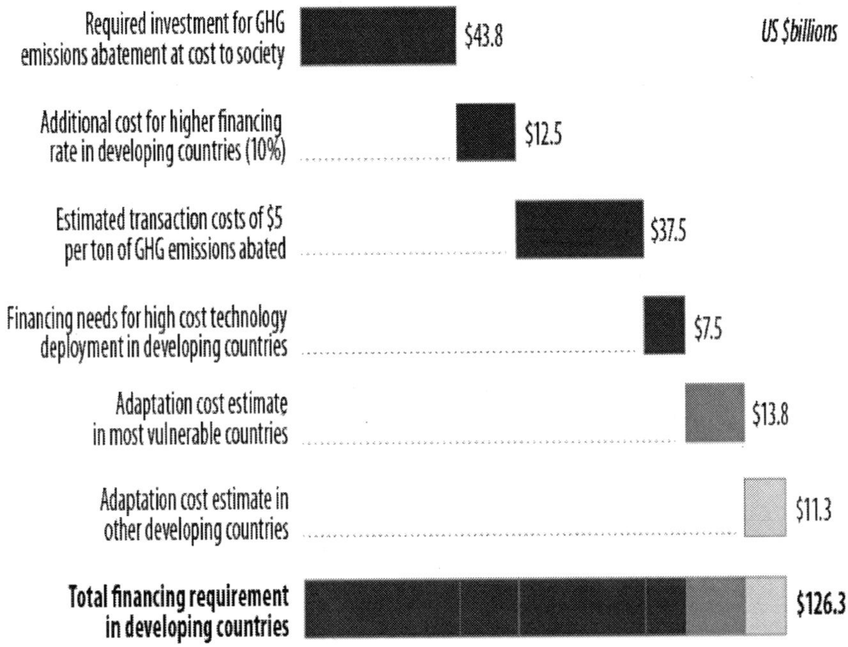

Source: Project Catalyst, *Toward a Global Climate Agreement: Synthesis Briefing Paper*, June 2009.

Figure 4. One Estimate of Annual Climate Change Financing Needed for Mitigation and Adaptation Activities in Lower Income Countries, 2010-2020 In Constant 2008 US$ Billions.

Further, with lower per capita incomes, weaker institutions, and limited access to technology, lower income countries are considered to have less adaptive capacity. See Figure 3 for one analysis of vulnerability and social capacity by country and region. The figure shows that in many cases a lower social capacity (as measured by literacy, education, health, and governance indexes) may make countries more vulnerable to the impacts of climate change.

Estimating the costs of adaptation with precision is difficult, not only because adaptation measures are widespread and heterogeneous, but also because the measures are embedded in the broader network of economic development strategies. While any investment in education, health, sanitation, and security, for example, may constitute good development, it also may help reduce socioeconomic vulnerability to both climactic and non-climactic stress factors. A variety of studies have tried to estimate the incremental costs of adaptation in low income countries. Methods, definitions, and scopes of adaptation in these studies vary, accounting for many of the differences in cost estimates.[36] In particular, many estimates assume that some portion of the incremental costs will be covered by the recipient countries themselves. Some studies attempt to consider "all" costs of adaptation to climate change and resulting damages (although none are comprehensive); some include just large-scale adaptation costs (i.e., not private measures taken by individuals); and some try to discern just the need for public financing for adaptation. The UNFCCC Secretariat estimates that the additional annual investment and financial flows needed worldwide would be on the order of $49 billion to $171 billion by 2030, with $30 billion to $73 billion needed for lower income countries (the largest element of uncertainty in the UNFCCC estimate lies in the cost of infrastructure adaptation). Other sources have produced varying estimates. See Table 1 above for adaptation cost estimates from a variety of studies.

Total Costs for Mitigation and Adaptation Activities in Lower Income Countries

A variety of international institutions and non-governmental organizations have used various climate change mitigation and adaptation analyses to estimate the climate-related financing needs for lower income countries. One example by Project Catalyst (Figure 4) sums the incremental costs for the period 2010-2020 to give an average total of $45 billion per year for mitigation needs in lower income countries. Factoring in a higher rate of investment in lower income countries and covering transaction costs and specific funding for emerging technologies brings the total financing requirement for abatement in lower income countries to around $100 billion annually. Including projections for the additional costs of adaptation activities gives a final estimate of $126 billion per year for the period 2010-2020 in lower income countries.[37]

SOURCES FOR INTERNATIONAL CLIMATE CHANGE FINANCIAL ASSISTANCE

As outlined in the previous section, trillions of dollars may be advocated over the coming decades to provide scaled-up, new, additional, predictable, and adequate[38] financing for lower income countries to enable and support their actions on climate change. These investments would aim to upgrade and expand energy, industry, and transport infrastructure; to manage land use, land use changes, and forestry practices; and to support the implementation of adaptation activities for reducing climate vulnerability and building climate resilience. Having estimated the potential financial costs for climate change investment, the next step in the process would be to consider the sources from which these funds may be generated. While markets that are privately constituted and self-regulated have delivered moderately to climate change investments worldwide, public institutions—including national governments, international organizations, and official financing mechanisms of the UNFCCC[39]—continue to be key drivers for climate change investments, specifically in lower income countries. In the past, these institutions have relied heavily on government revenues to finance their activities. But, with climate-related costs rising into the tens or hundreds of billions of dollars a year, it is unlikely that direct budget contributions from governments can meet the demand. Other sources of finance may be sought, and many proposals exist.

Potential sources of international climate change financing for lower income countries can be divided into five broad categories: (1) private sector; (2) public interventions to stimulate private sector investment; (3) public sector; (4) innovative finance; and (5) voluntary actions. Most potential sources of finance would require some measure of government action or oversight— either directly, through budget contributions or a transfer of funds, or indirectly, through state- sponsored regulations or other incentives to leverage private sector investment. See Appendix A for an extended glossary of terms related to the various sources of climate finance discussed in this section.

Each of the categories identified above could potentially generate funds to address climate change in lower income countries. Each has advantages and disadvantages. There is no single set of criteria for comparing these options.[40] Some of the criteria employed by commentators include the potential magnitude of funds that could be generated by each source; the economic and/or the GHG-related efficiency of each source; the practicality and predictability of generating funds from each source; the plausibility of assessing the "additionality"[41] of each source; the overall accessibility and transparency of the funds provided and their use; and the equity and incidence effects as expressed among countries or between higher and lower income countries. A brief summary of some of the more significant outstanding issues regarding the choice of sources is included in a discussion at the end of this section "Caveats Regarding Sources"; and a tabulated comparison of the various sources is offered in Appendix B.

Private Sector Sources

Private capital markets can provide one source for mobilizing financing for low-emission investments in lower income countries. Instruments such as foreign direct investment, portfolio investment, microfinance, and public-private partnerships could be promoted to scale up private financing for climate change mitigation and adaptation activities. Incentive structures may need to shift in order to favor such investment, and economic and/or regulatory policies may need to be implemented to define targets and raise the profitability of alternatives. The section below outlines possible sources for international climate change financing in a privately constituted and self-regulated market. Public sector interventions to stimulate private capital investments are addressed in the next section.

Foreign Direct Investment: Foreign direct investment is the long-term participatory investment in the ownership of productive assets—such as factories, mines and land—by a multinational corporation in a developing country's energy, industry, or transport sector. It can be a relatively stable source of financing. It has the greatest advantages in mitigation activities, in terms of transferring technology and standards which could allow economic development to leapfrog into more climate-friendly sectors such as energy efficiency and renewable energy. Many high- emission sectors—such as road transport, metals, mining, chemicals, timber, and cement—are dominated by large multinational corporations, and their investments and practices may likely have a big influence on the timing of alternative development pathways in lower income countries.[42] However, given that foreign direct investment tends to lag rather than lead economic growth, it is unlikely to play a significant role in the early stages of a shift onto such a development pathway, particularly given the initial high degree of uncertainty and the absence of the in-country inputs that large international firms need in order to operate efficiently.[43]

Portfolio Investment: Portfolio investment is the purchase of stocks, bonds, and money market instruments by foreigners for the purpose of realizing a financial return but not resulting in foreign management, ownership, or legal control. It could also be a stable source of international climate change financing. Investment could be mobilized through venture capital funds or specific "green" funds. It could appeal to investors willing to allocate investments to options that might generate less return but have greater potential in terms of climate change mitigation and socially responsible business practices. Currently, almost all "green" investment opportunities are concentrated in the more industrialized countries or the countries with emerging economies. Funds made available through this channel to lower income countries have been both limited and skewed in favor of one or two countries. Without other incentives, the amount of resources that can be raised is likely to remain quite small.[44]

Microfinance: Microfinance is the provision of financial services, in the fora of small loans at market value, to lower income country clients who traditionally lack access to banking and related services. It could serve as another vehicle for mobilizing local private resources for investments in climate-friendly development. Over the past three decades, microfinance has grown dramatically, with more than 7,000 microcredit institutions in 2006, serving about 80 million people in about 65 countries.[45] Climate-relevant microfinance has expanded beyond merely encompassing programs of credit provisioning to include schemes of microsavings and microinsurance. Given the close links between poverty reduction and climate vulnerability, scaling up microfinance has been considered a possible source of finance for climate adaptation programs. However, observers note that scaling up microfinance for long-term investment in productive activities and sustainable development would require support through a broader development strategy, including investments in infrastructure and human capital.[46]

Public-Private Partnerships: Public-private partnerships are business ventures funded and operated through a partnership between government and one or more private companies. They involve a contract between a public-sector authority and a private party, in which the private party provides a public service or project and assumes substantial financial, technical and operational risk in the project. Public-private partnerships have helped stimulate private financing for energy efficiency and renewable energy projects in lower income countries. Likewise, they have assumed a growing importance as a vehicle for financing climate change adaptation programs, in the form of infrastructure projects and the delivery of health services.[47]

Public Interventions to Stimulate Private Sector Investment

Much of the economic policy debate on climate change has been dominated by the search for market-based solutions to the problem of market failure (i.e., the external costs of GHG pollution).[48] Successful policy would address the externalities of GHG pollution in the market and reveal the costs of choosing high-emissions over low emissions technologies.

Once determined, these costs could be internalized through economic incentives (e.g., setting emissions levels or compliance pricing, etc.) to help drive pollution abatement. Funds for public or private sector contributions to international climate change financial assistance for lower income countries could be derived and transferred in any number of ways from the finances generated by the market-mechanism.[49]

In general, market-based mechanisms to reduce GHG emissions specify either the acceptable emissions level (quantity) or the compliance costs (price) and allow the marketplace to determine the economically efficient solution for the other variable. For example, tradable permit programs set the amount of emissions allowable under the program (i.e., the number of permits available limits, or "caps," allowable emissions), while allowing the marketplace to determine what each permit will be worth. Likewise, carbon fees set the maximum unit cost (per ton of CO_2e) that one should pay for reducing emissions. Private decisions would determine how or how much pollution actually gets reduced. In one sense, preference for a carbon fee or a tradable permit system depends on how one views the uncertainty of costs involved and benefits to be received.[53]

Tradable Permit Systems: Tradable permit systems—better known as "cap and trade programs"—set an overall cap on GHG emissions and then issue tradable permits to firms which would allow them to emit specified quantities of GHG.[54] Firms are required to hold a number of permits equivalent to their level of pollution. Those firms that need to increase their permissible quantities would purchase permits on the market. Those that could reduce their emissions more cheaply could potentially sell their allowances on the market. Some cap and trade programs would auction all permits during the initial issuance. Some programs would phase in permits through a period of free or reduced price allocations of permits. Some programs include a mechanism whereby firms could gain emission credits through pollution reductions performed in economic sectors outside of the regulated market (e.g., agriculture, forestry, or in countries or regions not covered by the program).

GHG Emissions and Economic Externalities

In economic theory, GHG emissions are considered a form of pollution and are thus characterized as "negative externalities."

A negative externality (also referred to as a negative transaction spillover or an external cost) is defined as the cost of a transaction incurred by an individual who did not participate in the transaction; and, because this cost is incurred outside the transaction, it is not reflected in the market price. For example, buying fossil-fuel generated electricity may lead to air pollution and adverse health effects, the costs of which (both monetary and non-monetary, such as illness, suffering, or curtailed activities) are carried by many people, not necessarily just the electricity producers and con-sumers within the transaction. In most cases, the costs of these third-party illnesses, suffering, or curtailed activities have not been factored into the market price of the electricity.

For many economists, GHG emissions are an acute example of a negative externality for several reasons. First, GHG emissions are highly diffuse and dispersed—GHG pollution involves individuals and firms in every country and continent on the planet, and each polluter makes only a very small contribution to GHG pollution. This makes GHG emissions very difficult to control (contrast this with "point pollution" where pollution emerges from one fixed, well- defined, and easily identified and controllable source, like a factory polluting a river). Second, the environ-mental costs of GHG emissions are also highly dispersed, as they are borne by individuals and firms in every country and continent on the planet. Third, incurring these risks is completely disconnected from responsibility for emitting GHG, and there are no simple institutional (political, social, and economic) or technological ways of linking them.[50]

Under the conditions of externalities, a free market (i.e., a privately constituted and self-regulated market) is "inefficient." There is no incentive to limit pollution because the full cost of polluting the atmosphere with GHG emissions is not borne by the person or firm or country responsible for the pollution. For the market to be efficient, producers and consumers would need to internalize the external costs of pollution into the price of their transactions (i.e., determine a "price on carbon"). Theoretically, market efficiency would be achieved when the price of GHG emissions is set to equal the marginal benefits of abatement.[51] Policymakers differ on the mechanisms through which to internalize the external cost of pollution, but many see some mixture of public sector regulation and/or market based incentives as the primary policy instruments.[52]

a. Revenues from domestic auctioning of emission allowances in domestic emission trading schemes: This would involve auctioning of domestic credits (as in the EU Emission Trading Scheme phase III, or any potential domestic cap and trade program) and transferring some part of associated revenues to international climate change financing.
b. Revenues from international auctioning of emission allowances in international emission trading schemes (such as Assigned Amount Units under the Kyoto Protocol): An Assigned Amount Unit (AAU) is a tradable "carbon credit" representing an allowance to emit GHG under the UNFCCC Kyoto Protocol emissions trading market. AAUs are issued up to the level of initial "assigned amount" of an Annex 1 Party. Some countries support sourcing revenue for international climate change activities by retaining some allowances from the Annex I countries and then auctioning them.[55]
c. Revenues from offset levies: This would involve withholding a share of offset revenues from emissions trading markets—such as the Kyoto market or any domestic market—as an international source of climate change financing. This is currently done with the Kyoto Protocol's Clean Development Mechanism (CDM),[56] which capitalizes the UNFCCC Adaptation Fund through a 2% levy on the proceeds from its certified emission reductions.[57]

Emission-Based Fees, and Other Levies: An emission-based fee is an environmental fee that is levied on the GHG content of fuels or other sources of emissions. By increasing the cost of emissions, emission fees raise public revenue, any or all of which could be transferred and used for international climate change assistance to lower income countries. Emission fees are occasionally referred to as "carbon taxes"; under guidelines of the U.S. Office of Management and Budget, a "tax" is primarily for generating revenues, while a user (i.e., emissions) fee[58] is primarily to charge for an entity's use of a resource (e.g., the atmosphere as a place to discharge waste emissions).[59]

a. Revenues from GHG fees: This would involve a tax on GHG emissions in countries raised on a per ton emitted basis.
b. Revenues generated from taxes on international aviation and shipping: This would involve either a levy on maritime bunker/aviation jet fuels for international voyages, or a levy on passenger tickets of international flights.

c. Revenues generated by removing fossil energy subsidies: This would involve public funds made available by the removal of fossil energy subsidies which could be diverted towards international climate change financing. While not a levy per se, redirection of subsidy grants or increases in tax receipts by reducing credits and deductions would function in a comparable economic manner to an emission tax on consumers.
d. Revenues from fossil fuel extraction royalties/licenses: This would involve a redirection of a portion of existing government receipts associated with domestic fossil fuel production for use in international climate change financing.[60]

Public Sector Sources

Allocating resources in a national public budget directly to international climate change assistance is a straightforward way for governments to finance activities in lower income countries, and historically public funding has played an important role in both mitigation and adaptation financing. In practice, public funds may be mobilized similar to, or as part of, official development assistance (ODA); or, public funds may flow through international financial institutions as grants or grant-equivalent (i.e. "concessional") loans.[61] While some see public funds as a practical, equitable, and potentially predictable source of international climate change financing, political acceptability in the donor countries over the longer term may depend on national circumstances and on the size of the contribution. Further, global fiscal cycles can place public finances in many high income countries under extreme pressures and could make it difficult to generate sufficient and reliable financial flows over the required period or in the required order of magnitude.[62]

Voluntary Budget Contributions: Voluntary contributions involve public revenues provided to recipient countries—either directly or by international financial institutions—through national budgetary decisions. Voluntary contributions have played—and likely may continue to play—the most important role in publically-funded international climate change finance. Voluntary contributions may draw revenue from a domestic base (e.g., through taxes or fees) and allow for contributing governments to (1) pursue different options at different times as public opinion evolves, (2) divert only a portion of revenue from a particular source for international climate change financing, rather than the entire revenue flow, and (3) retain control over annual spending, rather than provide some kind of automated mechanism. But voluntary contributions in many countries are subject to legislative decision making and annual appropriations, making predictability and reliability difficult to ascertain.

Mandatory or Assessed Budget Contributions: Some lower income countries[63] have proposed that UNFCCC Annex I parties contribute from 0.5%-1 .0% of their gross national income to climate change financing in non-Annex I countries, to be channeled through a multilateral fund under the authority of the Convention. This would generate approximately $ 150 billion to $300 billion per year at pre-crisis income levels of major Organization for Economic Cooperation and Development (OECD) economies. Others have proposed assessed contributions formulated on the basis of some combination of a contributing country's GHG emissions, population, and gross domestic product, in accordance with the principle of common but differentiated responsibilities and respective capabilities. Supporters propose that mechanisms should be put in place to make contributions legally binding. The source for these funds would be similar to voluntary contributions (i.e., domestic taxes or fees).

Public Debt Instruments: Public debt instruments raise money for public entities by borrowing from bond markets. Most high-income country governments can borrow money at a discount because their chance of default is considered low compared to privately held companies. Multilateral Development Banks (MDB) and other International Financial Institutions (IFI) use the same principle for raising capital for lower-income country govern-ments. Using their good credit rating, which is based on the fact that they are backed by the capital subscriptions of developed country governments, IF Is and MDBs borrow money at favorable conditions to lend at a lower interest rate or accept a higher risk, a benefit that they can pass on to their clients in the interest of development and climate protection. Increased capital subscriptions from higher income countries would allow these institutions to increase lending to lower income countries.

Innovative Finance

Despite the spectrum of sources available for international climate change financing, many observers consider current funding inadequate, and look to more innovative methods of finance. Various proposals have been made for raising revenues for climate action from sources not closely linked to GHG emissions. While these proposals have the capacity to generate large quantities of financial assistance, they may be perceived as arbitrary in their choices and by the numerous competing causes that could benefit from their financing. Two of the more heavily discussed proposals include a financial transaction tax and special drawing rights.

Financial Transaction Tax: A Financial Transaction Tax (FTT) is a levy on international financial transactions. The level of estimated revenues from the FTT is driven by the tax base, the tax rate, and the elasticity of the transaction volume to the tax rate. A global FTT, as currently debated, would be a new and additional source for climate finance. Strong international coordination and allowances for international implementation could increase the efficiency of such a source. However, critics point out that FTTs are unconnected to GHG emissions in any practical way, and the unresolved issues of incidence on both higher and lower income countries would make it difficult to implement universally.

Special Drawing Rights: Special Drawing Rights (SDRs) are an accounting mechanism— sometimes called "virtual currency"—typically held as a reserve asset in financial organizations such as the International Monetary Fund (IMF). SDRs supplement IMF member countries' official reserves and generate liquidity in the event of balance of payments difficulties. The value of an SDR is based on a basket of key international currencies. In some proposals to capitalize a fund for climate change financing, SDRs could be issued in exchange for real currency to generate revenues. In other proposals, bonds could be issued on the back of paid-in SDRs to generate liquidity and contributors would receive an equity stake in the fund proportional to their contribution. In either case, the IMF would not necessarily be the entity issuing these proposed SDRs or managing the system. Most policymakers have not supported the use of SDRs to capitalize resources for climate change finance because the effort would undermine the primary purpose of the SDR system, present legal and political/financial challenges in implementation, and offer few if any advantages over traditional capitalization.

Voluntary Actions

Some sources for international climate change financing may be found in philanthropic or voluntary markets. Proponents note that some companies and some consumers have already begun to implement voluntary changes and have already begun to make voluntary contributions in order to reduce GHG emissions. Many believe that absent more aggressive governmental intervention, it is unlikely that these trends would be quantitatively sufficient and timely enough to make a significant impact on climate change. Voluntary action may also hurt relative competitiveness and increase costs in the short term, reducing incentives to adopt more stringent standards in the future.

Philanthropy: Many philanthropic organizations already provide contributions to climate change mitigation or adaptation financing in lower income countries. Many work closely with nongovernmental and civil society organizations to promote education, knowledge sharing, and human capital advance-ments to further climate change investment in lower income countries.

Voluntary Offsets: Voluntary offsets are "carbon neutral" certificates that are sold by some private entities in exchange for climate-related services (e.g., a contribution to an international climate change fund). Companies currently sell offsets in exchange for assurance that the funds will be used to reduce GHG emissions (e.g., by planting trees). This may differ from general philanthropy in the sense that funds are, in principle, directly in exchange for quantified emissions reduction performance and could be issued through aggregators of small, diversified projects or through brokers.

Caveats Regarding Sources

The Role of Markets: The economic debate within international climate change policy has been dominated by assessments of market-based mechanisms aimed at changing price incentives so that investment in low-emissions development becomes more attractive (e.g., cap and trade, carbon fees, loan guarantees). Many agree that private investment will likely have a predominant role to play in any low-emissions economic future, and that establishing a price on GHG emissions will likely have a part in any effective policy agenda. However, concerns remain whether such mechanisms can induce the required shifts in production and consumption patterns and mobilize the necessary investment. Some assert that price mechanisms are unreliable guides in cases where investments are large, where returns are not immediately visible, and where conditions are dependent upon unpredictable policy initiatives. The uncertainties in investments are heightened when the climate and development challenge takes place against a backdrop of systemic financial market failure and natural resource price volatility. As such, some policymakers believe that market mechanisms would contribute only a partial role in a larger package of measures that includes a reliance on regulations and large-scale public investments.[64]

The Role of Governments: Notwithstanding economic considerations, the "private funding" versus "public funding" debate also has political, legal, and equity components. For example, some recipient countries contend that donor governments should provide public funds as the main source of climate change assistance because they understand climate finance as an international equity issue, with contributions serving as reparation for past environmental loss or damage. Others assert that developed country governments—not private corporations—have signed onto legally binding international environmental agreements to provide assistance to lower income countries. Some may believe public monies would be more direct and easier to generate, and therefore more predictable and sustained. They may not consider or recognize the challenges in some countries to appropriating federal funds for international purposes. Similarly, some recipient countries may be suspicious of foreign private investment and would prefer the funding to be under the control of local governmental decision-makers, hoping this would better reflect local priorities and indigenous cultures. Conversely, donor countries tend to underscore the costs of extending such financing, including the direct outlays of funds, the secondary costs to their domestic economies for investing abroad at concessional terms, and the losses accrued by passing funds through governments institutions or other intermediaries.

The Requirement for Scaled Up, New and Additional, Predictable and Adequate Financing: Most, if not all, low income countries have stated that fulfilling their commitments under the UNFCCC would depend on financial and technical support from higher income countries. As noted above, they seek resources that can be defined as "scaled up, new and additional, predictable and adequate."[65] While the term "scaled up" presumes an increase in funding from existing sources, the terms "new," "additional," and "adequate" are subject to diverse interpretations and controversy. "New" funds could signify entirely unique funding sources arising from new public levies, new international allocations, or new multilateral mechanisms; or, it could simply refer to funds from a new fiscal year, a new multilateral replenishment contribution, or a new domestic or international program that takes the place of an expiring one. Some are concerned that funding is not shifted merely from one type of development assistance to climate change assistance, with little or no increase comparable to the stated needs. "Additional" is meant to denote an increment above and beyond "business-as-usual." However, speculating as to counter-factual development assistance trajectories that would have taken place if not for the "additional" funding is rife with debate. Finally, the term "adequate," with respect to needs, is a wholly subjective quantity.

METHODS FOR DELIVERING INTERNATIONAL CLIMATE CHANGE FINANCIAL ASSISTANCE

The previous section outlined many of the existing and proposed funding *sources* for investment in climate change mitigation and adaptation activities in lower income countries. The next step would be to consider the methods through which these funds could be transferred from contributing countries to their recipients. A variety of mechanisms, organizations, and institutions for disbursing international climate change financing already exists. All have a role in catalyzing climate action: mobilizing additional resources; reorienting public and private flows toward low- carbon and climate-resilient investments; supporting the research, development, and deployment of climate-friendly technologies; and strengthening the institutional capacities of recipient countries. Mechanisms can be divided into three broad categories: (1) private or quasi-private sector, (2) public sector bilateral, and (3) public sector multilateral.[66] Foreign Direct Investment (FDI), Export Credit markets, non-concessional lending at the Multilateral Development Banks (MDB), and the various Kyoto Protocol market mechanisms at the UNFCCC (e.g., the Clean Development Mechanism (CDM)) would be classified as private or quasi-private sector mechanisms. Public sectors mechanisms would include contributing countries' Official Development Assistance (ODA) as well as many of the multilateral environment and development trust funds (e.g., the Global Environment Facility (GEF)) and the concessional lending windows housed at the various institutions at the World Bank Group.

Figure 5 presents a comparison of the financial flows for energy and infrastructure development—including development specific to climate change mitigation—in low-income countries per annum (in this example, the analysis compares flows for the year 2007). The figure shows that total investment in all "mitigation-relevant" sectors (i.e., funding for economic development in all key sectors that shape future GHG emissions in developing countries, including energy, transport and water infrastructure, industry, waste management, agricultural, and forestry) amounted to an estimated $316 billion, of which over 80% was from private sector funds.

It should be noted that "mitigation-relevant" investments need not be GHG reducing investments (e.g., both wind power generation and fossil fuel power generation are "mitigation- relevant" investments).

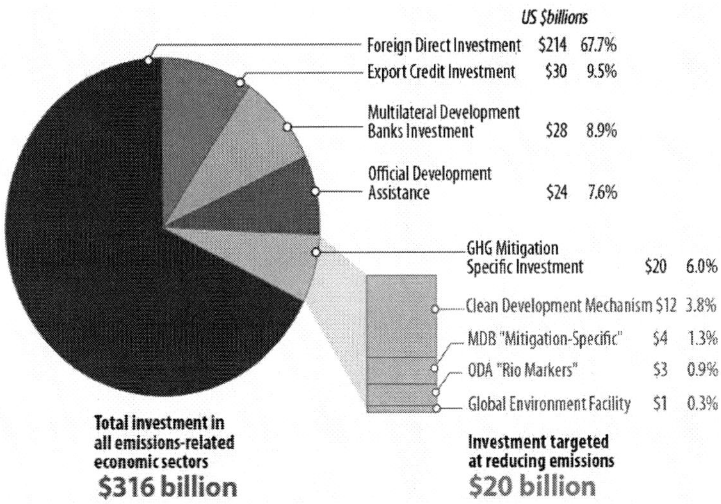

Source: Jan Corfee-Morlot et al., *Financing Climate Change Mitigation: Towards a Framework for Measurement, Reporting, and Verification*, Organization for Economic Cooperation and Development, October 2009.

Notes: "Total Investment in all emission-related economic sectors" refers to assistance for economic development in all key sectors that shape future emissions in developing countries, including energy, transport and water infrastructure, industry, waste management, agricultural, and forestry." "Emission-related" investment may serve to increase or decrease GHG emissions. "GHG mitigation-specific" assistance refers to assistance aimed "to achieve GHG mitigation in developing countries as its main objective." "ODA "Rio Markers" refers to Official Development Assistance contributed by OECD Members of the Development Assistance Committee (DAC), who report their aid activities to the OECD Creditor Reporting System (CRS) database, that specifically targets the objectives of the three United Nations Rio conventions (i.e., U.N. Convention on Biological Diversity, U.N. Framework Convention on Climate Change and U.N. Convention to Combat Desertification).

Figure 5. Total Investment in Emissions-Related Economic Sectors in 2007.

"Mitigation-specific" investments (i.e., investments in which the primary objective is to reduce GHG emissions) amounted to $20 billion in 2007—or 6% of total key sector investment—of which approximately 60% was from private sector funds.

All of the mechanisms identified above could potentially deliver funds to address climate change in lower income countries. Each has advantages and disadvantages, and there is no single set of criteria for comparing these options. Many critics contend that the overall architecture of financial mechanisms to address climate change is underfunded and unnecessarily complex. The array of funds and financial institutions lack both strategic mandate and adequate coordination, leaving many gaps, overlaps, and inefficiencies. Divisions have arisen over the proper financial instruments to employ in lower income countries (e.g., grants or loans) as well as the role shared by the public and private spheres. A brief summary of some of the more significant outstanding issues regarding the choice of sources is included in a discussion at the end of this section "Caveats Regarding Mechanisms."

Private Sector Mechanisms

The sheer magnitude of the required investment for climate change mitigation and adaptation activities necessitates capital flows from the private sector. Currently, the private sector accounts for over 85% of global investment in those economic sectors relevant to climate change mitigation and adaptation activities, although governments largely control the underlying infrastructure investments that affect the opportunities for low-emissions economic development. Similarly, public sector financing has been more prominent in low income countries—particularly the least developed countries and the small island states—in which private entities are still reticent to invest. To this point, private capital markets have filled in the gaps in climate change financing in lower income countries, and, in some cases, have taken the lead in market-ready mitigation investment to create a low-carbon economy, such as energy-efficient machinery, cleaner cars, and renewable energy.

Foreign Direct Investment: Over the past two decades, the international climate change agenda has shared a stage with an expanding global economy.[67] As such, particular attention has been paid to foreign direct investment (FDI) in lower income countries to address climate change. FDI has many potential benefits, including financing infrastructure expansion without contributing to public debt, supporting technology and knowledge transfer, and acting as a catalyst for further capital inflows. Despite considerable efforts to attract FDI in the last several years, actual levels of such investment into the energy and industry sectors in many countries with economies in transition have been moderate.[68] Similarly, FDI tends to lag rather than lead economic growth, and, as such, is not likely to play a significant role in the early stages of a shift to lower-emission development trajectories. Mobilization of the necessary capital resources requires an attractive investment climate—a business-friendly environment, favorable macroeconomic performance, and a regulatory environment that is predictable, fair, transparent and efficient.

Quasi-Private Sector Mechanisms

Several mechanisms function predominantly in the private sector but were set up initially by public sector entities or are currently backed by guarantees (whether financial or institutional) from them.

Export Credit Agencies: Export Credit Agencies (ECA) are private or quasi-governmental financial institutions or agencies that provide financing to domestic companies for their international trade activities. ECA services can include such instruments as direct loans, loan guarantees, and insurance for companies in order to help promote exports. These programs are implemented in cases where the private sector is unable or unwilling to provide financing to ensure equitable competition for U.S. exporters due to potential commercial, exchange rate, or political risks and uncertainties in overseas markets. The primary objective of ECAs is to remove the risk of repayment to exporters by shifting the financial burden of uncertainty onto themselves, for a premium. The U.S. Export-Import Bank and the Overseas Private Investment Corporation are two examples of export credit and overseas investment agencies connected to the United States government.[69]

Development Banks: Non-Concessional Lending**:** The International Bank for Reconstruction and Development (IBRD) and the International Finance Corporation (IFC) are the facilities within the World Bank Group that make non-concessional or "hard" loans to middle income countries as well as provide services for private sector ventures and projects in developing countries. While similar in structure to a commercial lending facility, the IBRD focuses primarily on investments that support poverty reduction, economic development, and global public goods, including food security and climate change. The IBRD currently has plans to increase lending to support renewable energy and energy efficiency projects in lower income countries, and it continues to serve as a center for research and knowledge-sharing on development practices, promoting lessons learned and identifying innovations to combat the effects of climate change. The regional banks in the World Bank Group have similar non-concessional lending windows. The United States supports the IRBD through its capital subscription to the IBRD General Capital Increase (GCI).[70]

UNFCCC Kyoto Protocol Mechanisms: The UNFCCC introduced three market-based mechanisms to assist countries with commitments under the Kyoto Protocol to limit or reduce greenhouse gas emissions. The mechanisms include International Emission Trading, the Clean Development Mechanism, and Joint Implementation. Most relevant to developing country assistance, the Clean Development Mechanism (CDM) allows entities with emission-reduction or emission-limitation commitments under the Kyoto Protocol to implement emission-reduction projects in lower income countries in order to earn saleable certified emission reduction (CER) credits which can be counted towards meeting Kyoto targets. Though the CDM has been used far less than many had initially envisioned (in part because of slow processes and governance issues), its Board says that it has issued more than 1.7 billion tons of CO_2e GHG reductions ($2.9 billion expected by end of 2012), and has leveraged US $33 billion from investors in 2007 alone.[71] Similar programs include the World Bank's Carbon Finance Unit, which uses donations from private and public entities to purchase GHG emission reductions in client countries.[72]

Public Sector Mechanisms

As a complement to private sector investment, many countries also contribute funds to climate change initiatives in lower income countries through various public sector mechanisms, organizations, and institutions. Governments may choose to contribute these funds either directly to recipient countries (i.e., bilateral assistance) or in combination with other donors through an international institution (i.e., multilateral assistance).[73] Bilateral assistance is often provided through a contributing country's development agency (e.g., U.S. Agency for International Development (USAID)). These direct and long-standing relationships between donor and recipient countries' development agencies can enable cooperative implementation plans with respect to environmental issues. Many climate change initiatives share a sensitivity to other development sectors (e.g., agriculture, biodiversity, health, and infrastructure) and provide an opportunity to implement innovative cross-sectoral programs. Further, bilateral assistance gives contributors more control over where the money goes and how the money is spent. For example, contributing countries may have more flexibility to allocate funds to countries that are of geopolitical strategic importance, but not facing the greatest development needs, than might be possible by providing assistance through a multilateral organization. By building a clear link between the contributing country and the recipient country, bilateral assistance may also garner more goodwill from the recipient country than if the funds had been provided through a multilateral organization.

Multilateral organizations offer different benefits for contributing countries. Multilateral organizations pool the resources of several contributors, allowing countries to share the cost of development projects (often called burden-sharing). In this way, one country's multilateral assistance is said to "leverage" additional funds from other contributing countries, as well as from implementing agencies, non-governmental organizations, the private sector, and even the recipient countries themselves. Further, long-standing and established multilateral institutions dedicated to climate change initiatives and sustainable development practices may hold a level of expertise and may benefit from knowledge carry-overs that are not as prevalent in the smaller bilateral assistance agencies of some contributing countries. Additionally, contributing countries may find it politically sensitive to attach or enforce policy reforms to bilateral assistance, and multilateral organizations can usefully serve as a shield for imposing and enforcing conditionality that may be politically sensitive to attach bilaterally. Finally, many believe that providing funds to multilateral organizations plays a role in a contributing country's leadership in the world economy.

Table 2 lists some of the most prominent bilateral and multilateral financing mechanisms for climate change activities in lower income countries. Below are descriptions of the various public sector mechanisms currently employed by the United States. Commentary on the effectiveness of each mechanism can be found in the topical discussion at the conclusion of this section as well as in the respective CRS reports footnoted under each heading.

Bilateral: Official Development Assistance[74]

Official Development Assistance: Official Development Assistance (ODA) programs, funded primarily through the U.S. Agency for International Development (USAID), aim to advance and sustain U.S. engagement with specific developing countries on critical global issues such as food security and climate change.

Table 2. Examples of Bilateral and Multilateral Financing Mechanisms for Climate Change Mitigation and Adaptation Activities in Lower Income Countries In Constant 2008 US$ Millions

Name	Funding Pledge (All Donors)	Use[a]	Details
Multilateral Funds Associated with the United Nations Framework Convention on Climate Change			
Global Environment Facility (GEF)-5	4,340	M	Provides grants. Time frame: 2010-2014, prior contributions from 1991 through 2009 amount to approximately $9 billion
Adaptation Fund	400-1,500	A	Provides grants. Time frame: 2008-2012; as of 2008, $91.3 million was available (4 million certified emission reductions (CERs) at €17.5 per CER)
Least Developed Countries' Fund	172	A	Provides grants. Includes pledges as of December 2008; $91.8 million has been received as of November 2008; operated by GEF
Sustainable Forest Management	154	M	Provides grants. Special program under GEF-4 for land use, land-use change and forestry
Special Climate Change Fund (SCCF Adaptation)	90	A	Provides grants. Include pledges as of December 2008; $68 million has been allocated to 15 projects as of November 2008; operated by GEF
Strategic Priority on Adaptation(SPA)	50	A	Pilot program on adaptation of the GEF Trust Fund
Green Climate Fund	—	A,M	Currently being negotiated under the UNF CCC Conference of Parties Ad Hoc Working Group on Long-Term Cooperative Action

Name	Funding Pledge (All Donors)	Use[a]	Details
Other Multilateral Funds			
Clean Technology Fund (World Bank Climate Investment Funds)	4,334	M	Time frame: 2009-2012; administered by the World Bank. Provides grants and loans; funded by the United States, the United Kingdom, Japan, among others
Strategic Climate Fund (World Bank Climate Investment Funds)	2,006	A,M	Time frame: 2009-2012; administered by the World Bank. Provides grants and loans; funded by the United States, the United Kingdom, Japan, among others
Forest Carbon Partnership Facility (World Bank)	300	M	Provides grants and loans; time frame 2008-2020
Global Facility for Disaster Reduction and Recovery (GFDRR)	84	A	Provides grants; time frame 2007-2010; targets high-risk low- and middle-income countries to mainstream disaster reduction in development strategies
United Nations Collaborative Program on Reducing Emissions from Deforestation and Forest Degradation in Developing Countries (UN-REDD)	35	M	Provides grants; administered by the UNDP; Norway, through its Climate and Forest Initiative, is the first donor, with US $12 million
Sustainable Energy and Climate Change Initiative (SECCI)	29	A,M	Provides grants and loans; the fund backs major investments in the development of biofuels, renewable energy, energy efficiency, and a wide range of sustainable energy options

Table 2. (Continued)

Name	Funding Pledge (All Donors)	Use[a]	Details
Bilateral Assistance			
Cool Earth Partnership (Japan)	10,000	A,M	Provides grants and loans; time frame: 2008-2012; up to US $2 billion to improve access to clean energy, and US $8 billion for preferential interest rate loans for mitigation projects
Climate and Forest Initiative (CFI)(Norway)	2,250	M	Provides grants; time frame: 2008-2012; pledged US $102 million to the Amazon Fund; other pledges to Brazil, Indonesia
International Window of the Environmental Transformation Fund (ETFIW) (UnitedKingdom)	1,182	A,M	Provides grants and loans; time frame: 2008-2010; most of the funds will be allocated trough the World Bank Climate Investment Funds
International Climate Initiative (ICI) (Germany)	764	A,M	Provides grants; funding for the initiative will be generated from auctioning 10 per cent of its allowances from the Emission Trading Scheme of the European Union (EU ETS); it has earmarked up to €120 million for the next five years
International Forest Carbon Initiative (IFCI) (Australia)	129	M	Provides grants; time frame 2007-2011; as of November 2008, US $50 million was allocated
United Nations Development Program-Spain MDG Achievement Fund – Environment and Climate Change thematic window	90	A,M	Provides grants; time frame: 2007-2010; Spain has pledged €528 to the Fund and US $90 million has been allocated for the Environment and Climate Change thematic window

Name	Funding Pledge (All Donors)	Use[a]	Details
Global Climate Change Alliance (GCCA) (European Commission)	76	A,M	Provides grants; time frame: 2007-2011; targets most vulnerable countries (least developed countries and small islands)

Source: Department of Economic and Social Affairs, *World Economic and Social Survey 2009, Promoting Development, Saving the Planet*, United Nations Publication, United Nations Publication, 2009, and World Bank, *World Development Report: 2010: Development and Climate Change*, and the websites of each listed fund.

a. Abbreviations: A, Adaptation; M, Mitigation; figures adjusted to 2008 US $.

ODA-funded programs are designed to support the efforts of recipient governments and their private sector and non-governmental partners to implement the political and economic changes needed for sustainable practice. USAID has targeted "adaptation," "clean energy," and "sustainable landscape" programs as the three pillars of the Obama Admin-istration's Global Climate Change Initiative to assist vulnerable countries adapt to the impacts of climate change and reduce net GHG emissions in their economies. USAID estimates that approximately 1/6 of its agency-wide Development Assistance account is earmarked for the Administration's Global Climate Change Initiative programs.

U.S. appropriations for bilateral climate change initiatives through USAID's or the Department of the State's ODA programs is subject to congressional approval. Congressional committees of jurisdiction include the U.S. House of Representatives Committee on Foreign Affairs, (various subcommittees); the U.S. House of Representatives Committee on Appropriations, Subcommittee on State, Foreign Operations, and Related Programs; the U.S. Senate Committee on Foreign Relations, Subcommittee on International Development and Foreign Assistance, Economic Affairs, and International Environmental Protection; and the U.S. Senate Committee on Appropriations, Subcommittee on State, Foreign Operations, and Related Programs.

ODA is a significant component of international climate change assistance. The Organization for Economic Cooperation and Development (OECD) has estimated bilateral, climate-specific support to low-income countries at an annual average of about $3.4 billion from 2003-2007, as reported in their Creditor Reporting System. This climate-specific financing represented about 0.0 1% of contributing countries' Gross Net Income (GNI) for that period, and about 3.4% of contributing countries' total bilateral ODA for all purposes. Accordingly, the United States' contributions represented about 0.002% of GNI and about 0.1% of all bilateral ODA, in both instances below the international average. Given the vagaries of definitions and reporting, the OECD estimates that all bilateral financing support for climate mitigation represented about US $8 billion to $53 billion in 2007—no more than 1/6 of the total estimated flows of about US $314 billion going to the sectors relevant to climate mitigation actions (i.e., energy, transportation, agriculture, water supply, industry, minerals, and mining).[75]

Multilateral: International Financial Institutions

International Financial Institutions: International Financial Institutions (IFI) are the primary multilateral mechanisms for climate change financing to lower income countries. IF Is coordinate multiple donor country contributions and provide loans, grants, and investment services to developing economies to promote growth, alleviate poverty, and aid in targeted programs such as climate change and food security. IF Is were designed to provide professional advice and technical support to address the economic impediments to developing country investment. In the United States, climate-related funding for IF Is is managed primarily through the Office of International Affairs at the U.S. Department of the Treasury.

U.S. appropriations for multilateral climate change initiatives through authorized International Financial Institutions are subject to congressional approval. Congressional committees of jurisdiction include the U.S. House of Representatives Committee on Financial Services, Subcommittee on International Monetary Policy and Trade; the U.S. House of Representatives Committee on Appropriations, Subcommittee on State, Foreign Operations, and Related Programs; the U.S. Senate Committee on Foreign Relations, Subcommittee on International Development and Foreign Assistance, Economic Affairs, and International Environmental Protection; and the U.S. Senate Committee on Appropriations, Subcommittee on State, Foreign Operations, and Related Programs.

a. Development Banks: Concessional Lending: The International Development Association (IDA) is a facility within the World Bank Group that makes grants and highly concessional or "soft" loans to the world's 79 poorest countries. IDA remains the single largest source of development finance globally across a range of sectors, including climate change mitigation and adaption initiatives as well as climate-relevant programs in primary education, basic health services, clean water and sanitation, environmental safeguards, business improvements, infrastructure, and institutional reforms. The United States was a driving force behind the creation of IDA in 1960 and remains its largest shareholder. The United States currently supports IDA through appropriated contributions pledged to IDA's 16th replenishment period.[76]

b. Tropical Forests Conservation Act: The Tropical Forest Conservation Act authorizes debt relief for low and middle income countries to support conservation of tropical forests. Under the program, treated debt is reduced and redirected to provide for grants to local non-governmental organizations and other entities engaged in forest conservation in the recipient country. The United States uses appropriated funds to pay for the budget cost of the debt restructuring. "Debt-for-nature" initiatives like the TFCA may be structured as either bilateral or multilateral programs and may employ third-party non-governmental organizations as brokers or rely on direct government-to-government agency coordination. To date, the United States has concluded 17 TFCA agreements in 14 countries, generating over $260 million for tropical forest conservation and the consequent GHG emission savings.[77]

c. Global Environment Facility: The Global Environment Facility (GEF) is an independent and international financial institution that provides grant-based financing to cover the additional or "incremental" costs associated with transforming projects with national development benefits into ones with global environmental benefits. GEF partners with international institutions, nongovernmental organizations, and the private sector to assist lower income countries with environmental projects related to six areas: biodiversity, climate change, international waters, the stratospheric ozone layer, land degradation, and persistent organic pollutants. Since its inception, GEF has allocated $9.2 billion—supplemented by more than $40 billion in co-financing—for more than 2,700 projects in 165 countries. GEF estimates that approximately 50% of its implemented projects assist climate change activities. The United States currently supports GEF through appropriated contributions pledged to GEF's fifth replenishment period (2010-2014).[78]

d. Clean Technology Fund: The Clean Technology Fund (CTF) is one of two Climate Investment Funds administered by the World Bank Group that aim to help finance lower income countries' transitions toward low-carbon and climate- resilient development. Implemented in 2008 by the United States, the United Kingdom, and Japan, and joined by several other donors in the international community, the CTF seeks to provide financing—principally to larger emerging economies and to regional groups—for demonstrating, deploying, and diffusing large-scale clean energy investments with the potential for long-term avoidance of GHG emissions. The fund promotes renewable energy and energy efficient technologies as well as energy efficiency strategies in the transportation, building, industry, and agricultural sectors. Eight contributing countries have pledged $4.4 billion to the CTF since September 2008 in support of programs in 18 lower income countries. The U.S. pledge of $2 billion—split between the CTF and the SCF—is currently supported through annual appropriated contributions.[79]

e. Strategic Climate Fund: The Strategic Climate Fund (SCF), founded in conjunction with the CTF as the second of the World Bank's two Climate Investment Funds, supports three programs that aim to pilot new and scaled-up approaches to address climate change challenges in lower income countries. The Pilot Program for Climate Resilience assists many of the poorest and most vulnerable countries prepare for and respond to the impacts of climate change by integrating climate adaptation measures into core development planning. The Forest Investment Program works to reduce deforestation in lower income countries through improved forest management. The Program for Scaling-Up Renewable Energy in Low Income Countries supports a select number of the poorest countries in their efforts to expand energy access and stimulate economic growth through the deployment of renewable energy solutions. Ten contributing countries have pledged $2.5 billion to the SCF since September 2008 in support of programs in over 30 lower income countries. The U.S. pledge of $2 billion— split between the CTF and the SCF—is currently supported through annual appropriated contributions.[80]

f. Green Climate Fund: The Green Climate Fund, currently under negotiation in the UNFCCC Ad Hoc Working Group on Long Term Cooperative Action as set forth by the Cancun Agreements in December 2010, is to be designated as the official financial mechanism of the Convention. Upon creation, it would support projects, programs, policies, and other activities in lower income country using thematic funding windows for mitigation, adaptation, forestry, capacity building, and technology transfer, among others. All program attributes have yet to be agreed upon by negotiating Parties, including provisions for its trustee, governance structure, administration, funding levels, and its official status under the Convention.[81]

Caveats Regarding Mechanisms

The Debate between Bilateral and Multilateral: Contrasting models of foreign assistance have arisen through the years in programs like the Marshall Plan, involving bilateral aid arrangements between countries, and the Bretton Woods process, involving multilateral arrangements like the IMF, the World Bank, and others. Historically, bilateral assistance has dominated the foreign aid landscape, and it remains potentially the most direct and timely method of reaching a destination and working for a recipient government. Environmental contributions by multilateral organizations have been criticized for a lack of cost efficiency, a focus on economic development in lieu of the environment, and the imposition of requirements for global environmental benefits as opposed to local, environmentally-sound development. But proponents stress that multilateral assistance tends to be less tied to the political self-interest of individual donor countries, allows for the efficient pooling of financial resources and the leveraging of additional monies, helps ensure that different bilateral arrangements do not work at cross-purposes, and serves to develop a sense of cooperation among nations with the additional advantage of reducing conflict.

The Complex Institutional Architecture: Whether it is the United Nations family of agencies, the World Bank Group, or any number of countries and their development agencies, the plethora of actors engaged in climate financing is unwieldy to many recipient countries. Larger bureaucracies tend to slow and dampen performance, reduce flexibility and transparency, and heighten transaction costs. As mechanisms for climate finance proliferate, each new or special- purpose fund carries with it a set of challenges, including redesigned institutional and governance functions, inefficient allocations, and limitations on scaling up. These challenges threaten to reduce the overall effectiveness of climate finance because as transaction costs increase, recipient country ownership lags and alignment with country development objectives becomes more difficult.

The Limitations of Current Market Mechanisms: The UNFCCC Clean Development Mechanism has demonstrated that markets can stimulate emission reductions, provide essential learning, and build capacity. But some attest that mechanisms such as the CDM contain inherent inefficiencies, including a relatively weak and inefficient UNFCCC governance structure, an inability to successfully generate local economic co-benefits, an uneven distribution of projects that is geared toward higher-income emerging economies, a weakness in incentives to foster real transformations to a low-carbon economy, and a debate over the additionality of a given project's reductions (i.e., a debate over if the enacted emission reductions paid for by the mechanism are additional to actions that otherwise would have occurred).

The Choice between Top-Down or Bottom-Up: Whether purposeful or not, the international community is currently operating under a disjointed set of mechanisms which encourages fragmentation of the global response to climate change. Some argue that the "bottom-up" approach offers competition for market share, flexibility in international commitments, and a greater leeway to manage climate change action on individual terms, including accounting for different views of national circumstances, domestic politics, legal backgrounds, economic costs, and competitiveness exposures. The United States and other higher income countries, as well as some emerging economies such as China, often stress the bottom-up approach as a way to protect international political issues (such as sovereign rights) and international economic issues (such as the globalization of markets). Others find fragmentation to be a great detriment to effectiveness, efficiency, and equity at a time when the international community may wish to bring together a myriad of elements into a single, functioning, strategic framework. Most lower income countries tend to support a centralized, top-down mechanism situated in the UNFCCC, or the United Nations at large. They may perceive a heightened sense of coordination, consistency, and transparency afforded by these institutions through their "one country: one vote" paradigm.

U.S. CONTRIBUTIONS TO INTERNATIONAL CLIMATE CHANGE FINANCIAL ASSISTANCE

Congressional Authority, Oversight, Appropriations

The United States has relied mostly on direct budget appropriations to finance climate change actions internationally. Congress is responsible for several activities in this regard, including (1) authorizing periodic appropriations for federal agency programs and multilateral fund contributions, (2) enacting those appropriations, and (3) overseeing U.S. interests in the programs.[82] Oversight may come in the form of guidance; please see box "Some Examples of Legislative and Executive Branch Guidance" in this section for examples. Currently, direct budget contributions from the U.S. government are appropriated to relevant federal agencies to support their bilateral and multilateral contributions to international climate change initiatives. These appropriations are requested on an annual basis by the Administration and enacted on an annual basis by the Congress.

The majority of bilateral and multilateral contributions to international climate change initiatives is funded through programs at the Department of State, the Department of the Treasury, and the U.S. Agency for International Development.[83] Funds for these programs are appropriated in the Administration's Executive Budget, Function 150, for State, Foreign Operations, and Related Programs. Historical budget authority for these programs through 2009 is presented in Table 3. The current Administration's budget authority and budget requests are presented in Table 4.

Congressional committees of jurisdiction for international climate change programs at the Department of State, the Department of the Treasury, and the U.S. Agency for International Development include

- the U.S. House of Representatives Committee on Foreign Affairs, (various subcommittees);
- the U.S. House of Representatives Committee on Financial Services, Subcommittee on International Monetary Policy and Trade;
- the U.S. House of Representatives Committee on Appropriations, Subcommittee on State, Foreign Operations, and Related Programs;
- the U.S. Senate Committee on Foreign Relations, Subcommittee on International Development and Foreign Assistance, Economic Affairs, and International Environmental Protection; and

- the U.S. Senate Committee on Appropriations, Subcommittee on State, Foreign Operations, and Related Programs.

Currently, no federally legislated system exists to require or assure future, predictable public financing for international climate change assistance (e.g., a mandated allocation of redirected fossil fuel subsidies). Similarly, no federally legislated market-based mechanism has been instituted that would contribute funds generated from the private sector for international climate change assistance (e.g., a GHG cap-and-trade program, which could incentivize international private capital investment through offset markets or provide internationally-targeted funds through an allocation of auction or offset revenue).

Historical Contributions

U.S. participation in international negotiations for global environmental assistance stretches at least back to the 1972 United Nations Conference on the Human Environment (UNCHE), the first concerted effort by the international community to focus on the environment as a major topic of concern and attention.[84] Therein, discussions among representatives from both industrialized and lower income countries gave rise to a compromise on a doctrine to address both environmental and developmental policy. Governments at the time agreed that (1) the environment and development are two mutually reinforcing sides of the same coin, and (2) that industrialized countries would accept the principle of "incrementality" by which they would pay some or all of the additional costs of environmental initiatives in the developing world above and beyond the basic costs of development. The relationship between the environment and development was further developed at the 1992 United Nations Conference on Environment and Development, in the document called *Agenda 21*,[85] wherein a blueprint was detailed for putting sustainable development into practice. *Agenda 21* also doubled the pledge by the industrial world to assist lower income countries in poverty alleviation and environ mental protection. Although *Agenda 21* was impressive in its scope and comprehensiveness, the policy was non-binding; and some observers note that "the whole enterprise [was] heavily dependent on strong leadership from major countries, adequate financing, and effective institutional arrangements for follow-up ... none of which materialized."[86]

SOME EXAMPLES OF LEGISLATIVE AND EXECUTIVE BRANCH GUIDANCE

The United States government occasionally offers policy guidance or attaches mandates to legislation regarding international climate change assistance. Guidance may be directed toward the recipient countries, the institutions who serve to implement the projects and programs, or U.S. officials who instruct in the multilateral fora. Guidance could be issued to support low-carbon transformations in institutional energy and environmental strategies; emission abatement measurement, verification, and reporting practices; institutional fiduciary standards, procurement practices, and environmental and social safeguards; among others. Some examples of mandates in federal orders and legislation include:

- The Omnibus Appropriations Act, 2009 (P.L. 111-8), which permits up to $10 million for the Least Developed Countries Fund, under the UNFCCC, to support grants for climate change adaptation programs. To receive the funds, the Global Environment Facility (GEF) must annually report on the criteria it uses to select programs and activities that receive funds, how funded activities meet such criteria, the extent of local involvement in these activities, the amount of funds provided, and the results achieved.
- The 2009 U.S. Department of the Treasury Guidance to MDBs for Engaging with Developing Countries on Coal- Fired Power Generation, which provides guidance to the senior management and covers a range of issues including alternatives analysis, power sector policy reform, and capacity building. The guidance is intended to be adapted by individual MDBs and incorporated into their respective operational policies, country and sector strategies, and other procedures that are related to the public or private project cycle for coal-powered generation operations.
- The International Development and Finance Act of 1989 (P.L. 10 1-240), which requires the U.S. executive directors of the MDBs to abstain from voting, or to vote no, on any project likely to have significant environmental impact if it has not been properly assessed, or if the assessments have not been made public at least 120 days before a vote.

Financial pledges were made to support climate change initiatives specifically under the 1992 United Nations Convention on Climate Change (UNFCCC). The United States, as a Party to the UNFCCC, has committed to providing financial and technical assistance to lower income countries' efforts in meeting their respective obligations. Historically, some of this assistance has flowed through the Global Environment Facility—the official financial mechanism of the UNFCCC—to which the United States has contributed funding since 1993. Further, at the 2009 UNFCCC Conference of Parties (COP) in Copenhagen, Denmark, and at the 2010 COP in Cancun, Mexico, the United States helped negotiate a package that included developed country pledges of an aggregated $30 billion of "fast start" climate financing for the period 2010 to 2012 and $100 billion annually by 2020. This funding is to come from a wide variety of sources, public and private, bilateral and multilateral, including alternative sources of finance, and be delivered through both new and existing fund arrangements, with governance structures providing for equal representation of developed and developing countries. These negotiations are still pending and are currently non-binding. The 2011 COP is scheduled to meet during December in Durban, South Africa to further negotiations toward a potentially binding treaty. The Obama Administration has not yet specified what share of pledges it envisions the United States providing, nor a strategy for how to fulfill the long-term pledge.[87]

Historically, the United States' credibility on international climate change financing has been impaired by periodic under-funding. The United States is currently $217 million in arrears for its assessed contributions to the Global Environment Facility (i.e., 10% of the total U.S. pledge of $2,185 million since 1994).

Also, though the Bush Administration helped establish the Clean Technology Fund in 2008 under the World Bank and pledged to help capitalize it, the U.S. Congress declined to appropriate the first U.S. payment of $400 million requested by the Administration for FY2009, and is far short of the $2 billion aggregated pledge the Bush Administration made for the period 2008-20 12. Further, recent appropriations fall well below what many countries envisage for the U.S. share of the $30 billion UNFCCC "fast start" pledge.[88]

Table 4 outlines recent U.S. budget authority and Administration requests for international climate change assistance programs by agency. For a more detailed description of the Obama Administration's Global Climate Change Initiative, its agencies, programs, funding, and purposes, please refer to Appendix C.

Table 3. Historical U.S Budget Authority for International Climate Change Assistance by Agency and Program Actual Funding, in Nominal US$ Millions

Agency/Program	'93	'94	'95	'96	'97	'98	'99	'00	'01	'02	'03	'04	'05	'06	'07	'08	'09
Department of State	**1**	**1**	**1**	**3**	**3**	**5**	**7**	**7**	**7**	**7**	**6**	**5**	**8**	**12**	**41**	**41**	**55**
Diplomatic and Consular Affairs															3	4	2
Economic Support Fund													1	6	32	32	24
International Organizations and Programs	1	1	1	3	3	5	7	7	7	7	6	5	7	6	6	5	29
Department of the Treasury	**0**	**12**	**35**	**14**	**14**	**18**	**60**	**14**	**54**	**43**	**56**	**52**	**44**	**46**	**46**	**46**	**46**
Debt Restructuring Tropical Forestry Conservation Act									13	5		20	20	20	20	20	20
Global Environment Facilitya		12	35	14	14	18	60	14	41	38	56	32	24	26	26	26	26
U.S. Agency for International Development	**200**	**173**	**192**	**175**	**147**	**163**	**236**	**156**	**157**	**173**	**208**	**193**	**183**	**190**	**99**	**116**	**222**
Andean Counter-drug Initiative											2	3	2				
Assistance for Eastern Europe and Baltic States							12	4	13	11	8	7	5	6	3	11	7
Assistance for the Independent States of the Former Soviet Union							35	34	31	30	48	47	34	30	5	15	8
Development Assistance	200	173	192	175	147	163	169	109	112	116	140	125	134	118	89	81	113
Development Credit Authority							1	1	1								
Economic Support Fund							19	8		12	6	9	5	33		7	94

Table 3. (Continued)

Agency/Program	'93	'94	'95	'96	'97	'98	'99	'00	'01	'02	'03	'04	'05	'06	'07	'08	'09
International Disaster Assistance										4	4	2	2	2	2	2	
P.L. 480 Food Aid													1	1			
Complementary Agencies[b]	*	*	*	*	*	*	*	*	*	*	*	*	*	*	*	*	*
Total	**201**	**186**	**228**	**192**	**164**	**186**	**303**	**177**	**218**	**223**	**270**	**250**	**235**	**248**	**186**	**203**	**323**

Source: U.S. Government Accountability Office, *Climate Change: Federal Reports on Climate Change Funding Should be Clearer and More Complete,* GAO-05-46 1, August 2005; Office of Management and Budget, Federal Climate Change Expenditures Report to Congress, (various years).

a.GEF funding as reported by OMB each year represents the portion of total GEF funding that is related to climate change.

b.International climate change assistance from "complementary agencies" may range from several millions of dollars to several hundred millions of dollars annually. Complementary agencies include, for example, the Department of Energy, Environmental Protection Agency, Department of Commerce, National Science Foundation, Department of Agriculture, National Aeronautics and Space Administration, Millennium Challenge Corporation, and U.S. Trade and Development Agency, and serve to supplement core international climate assistance. OMB did not report international climate change assistance from complementary agencies until 2010.

Prior to 20 1 0, an accounting of international climate change assistance contributed by agencies other than the core international assistance agencies (i.e., USAID, State, Treasury) can be interpreted from the various United States Action Reports, National Communication of the United States of America Under the United Nations Framework Convention on Climate Change. (See the Department of State's website for a catalogue of these reports, at http://www.state The U.S. Action Reports to the UNFCCC describe programs, give large aggregate numbers in support of total U.S. international assistance, and define "direct U.S. support" for international climate change activities to include "official development assistance (ODA), official assistance (OA), grants from foundations and other philanthropic institutions, U.S. government backed project financing, NGO funds, foreign direct investment (FDI), and commercial sales from private industry." Thus, for example, the Third U.S. Action Report states that total direct funding from the U.S. public and private sectors for climate change activities in developing countries for the period 1997-2000 is $7.7 billion (p. 125), wherein total assistance from the core agencies as reported by OMB (see table above) is only $830 million. While most of the bilateral U.S. government supported activities listed in the U.S. Action Reports are funded primarily through USAID, there are many “interagency” initiatives that contribute to international climate change assistance, including the following (as reported in the Fifth U.S. Action Report, 2010): The U.S. Country Studies Program (1994 and 2001); Group on Earth Observations; Asia-Pacific Partnership on Clean Development and Climate; Methane to Markets Partnership; International Partnership for the Hydrogen Economy; Carbon Sequestration Leadership Forum; ITER; Generation IV International Forum; Global Nuclear Energy Partnership; U.S./China Energy and Environmental Technology Center; Clean Energy Technology Export Initiative; U.S. Climate Technology Cooperation Gateway; U.S. Clean Energy Initiative; Efficient Energy for Sustainable Development; Global Village Energy Partnership; United States-Asia Environmental Partnership; International Renewable Energy Program; Climate Technology Initiative; Renewable Energy and Energy Efficiency Partnership; Partnership for Clean Indoor Air; ENERGY STAR program; eeBuildings; Collaborative Labeling and Appliance Standards Program; Integrated Environmental Strategies program; Central America Greenhouse Gas Inventory Improvement Project; among others. Specific funding levels for these programs or agency allocations are not always specified.

As outlined above, the methodology for reporting taken by the U.S. Action Reports for the UNFCCC is vastly different than the methodology presented by OMB in their reports to Congress. Interpreting funding levels as they relate to international efforts to address climate change is challenged by several reporting issues. The Government Accountability Office (GAO) Climate Change: Federal Reports on Climate Change Funding Should be Clearer and More Complete, GAO-05-461, lists some of the challenges as: (1) levels of aggregation of the budget request, (2) changes in scope of what is reported, (3) changes in accounting methods over time, (4) lack of descriptions by agencies in their budget documentation, and (5) omissions of reporting of some arguably climate-related activities in the overall program. Some reported funding changes represent differences in dollar amounts for the same activities over time. Other changes are because the scope of the climate change programs has changed, or programs have been redefined to fall under climate change initiatives. Also, methods of accounting for some activities have changed over time, introducing inconsistencies across years. Further, the activities counted as relating to climate change often are below the budget account level; and while some agencies provide adequate detail—though dispersed throughout their budget documentation—others do not mention the “climate change” relevance in their documentation of certain accounts although portions of them are included in the climate change reporting. See CRS Report RL338 17, *Climate Change: Federal Program Funding and Tax Incentives*, by Jane A. Leggett, for further elaboration on this issue.

Table 4. Recent U.S Budget Authority for International Climate Change Assistance by Agency and Program (Obama Administration) in Nominal US$ Millions

Agency/Program	2010 Enacted	2011 Request	2011 Enacted	2012 Request
Department of State	**199**	**149**	**TBD[a]**	**143**
Least Developed Country Fund	30	30	TBD	30
Special Climate Change Fund	20	20	TBD	20
World Bank Forest Carbon Partnership	10	15	TBD	21
Other State Programs	139	84	TBD	72
Department of the Treasury	**421**	**744**	**TBD**	**677**
Tropical Forests Conservation Act	20	20	TBD	15
Global Environment Facility[b]	26	89	46	72
Clean Technology Fund	300	400	185	400
Strategic Climate Fund: Pilot Program for Climate Resilience	55	90	/50[c]	40
Strategic Climate Fund: Forest Investment Program	20	95	/50	130
Strategic Climate Fund: Scaling-Up Renewable Energy	0	50	/50	20
U.S. Agency for International Development	**383**	**491**	**TBD**	**509**
Bilateral and Regional Programs	383	491	TBD	509
Complementary Agencies[d]	**77**	**110**	**TBD**	**TBD**
Total	**1080**	**1494**	**TBD**	**TBD**

Source: Office of Management and Budget, *Federal Climate Change Expenditures Report to Congress*, 2010; Office of Management and Budget, *The Budget of the United States Government*, 2011 and 2012 (State, Foreign Operations, and Related Programs Budget). H.R. 1473, the Department of Defense and Full-Year Continuing Appropriations Act, 2011 (P.L. 112-10).

a. TBD, "to be determined": Appropriated funds for these specific programs/activities are drawn from larger line item categories in agency budget authorities, occasionally with "shall"-language implementing spending ceilings. Based on provisions in H.R. 1473, allocations in FY20 1 1 for these programs are left at the discretion of the agency and have yet to be determined/reported.

b. GEF funding as reported by OMB each year represents the portion of total GEF funding that is related to climate change. Total GEF request across all environmental sectors is FY20 10 en. $87million, FY20 11 req. $175million, FY20 1 1 en. $90million, FY20 12 req. $144 million.

c. H.R. 1473, The Department of Defense and Full-Year Continuing Appropriations Act, 2011 (P.L. 112-10), FY20 11 budget authority for the Strategic Climate Fund is $50million for all programming. Allocation of funds among the three sub-programs is at the discretion of Treasury and has yet to be determined/reported.

d. Budget authority is for foreign assistance "core agencies (i.e., the Department of State, the Department of the Treasury, and the U.S. Agency for International Development) as presented in the State, Foreign Operations, and Related Programs account. Additional funds have been identified in the Administration's budget proposals for other "complementary" agencies, including the Department of Energy, Environmental Protection Agency, Department of Commerce, National Science Foundation, Department of Agriculture, National Aeronautics and Space Administration, Millennium Challenge Corporation, and U.S. Trade and Development Agency, to supplement core international climate assistance.

APPENDIX A. GLOSSARY OF OPTIONS FOR GENERATING AND DISBURSING FINANCING TO ADDRESS CLIMATE CHANGE

Table A-1. Glossary of Finance Options Explanations Are Neither Comprehensive nor Definitive of the Many Proposals That Exist

Fund Generation Mechanisms	
Private Compliance Market	Private sales and purchases of emission allowances, or credits for emission reductions, as under many Cap-and-Trade schemes, the Clean Development Mechanism of the UNFCCC, and other proposals.
Government Compliance Market	Purchases of emission reduction credits by governments from private entities or governments, such as through Joint Implementation under the Kyoto Protocol. Some European governments appropriate funds to acquire such credits, to be applied to meet the national Greenhouse Gas (GHG) target.
National Auctioning of Allowances	Designating for international finance a percentage of the proceeds of governmental auctioning emission allowances under national (or sub-national) emission control systems, including Cap-and-Trade.

Table A-1. (Continued)

Fund Generation Mechanisms	
Levy on Certified Emissions Reductions	A share of any certified emission reductions might be collected, to be sold or auctioned to generate revenues. Alternatively, a fee could be levied on issuance of certified emission reductions, proportionate to the quantity or at a fixed transaction cost.
Share of Proceeds on Emissions or Offset Trading	Collection of a percentage of the funding associated with sales of traded emission allowances or certified emission reductions (offsets), as part of registering the trade. This could happen in a domestic or international program.
Emissions Fees (Carbon Tax)	A fee levied on each unit of GHG emissions from sources.
Public Appropriations	Appropriations of funds for international finance (i.e., drawing on general purpose government revenues from income taxes, etc.)
International Auctioning of Allowances	Emission allowances or offsets from national programs could be transferred to an international or inter-governmental entity, which could then auction them internationally to generate funds.
Levy on Surplus or Banked Allowances	A fee on the transfer of unused allowances from one compliance period into a later one.
International Emi-ssions Allowances, with or without Trading, on Aviation and/or Maritime Transport	An international entity would be authorized to allocate or sell emissions permits to emission sources that are easily mobile across national boundaries, such as aviation and marine transport. This could be through an intergovernmental agreement among sovereign nations, not necessarily delegating any "governance" authority.
International Levy on Aviation or Marine Bunker Fuels	A tax could be levied on fuel use of emission-related entities, such as aviation bunker fuels or marine bunker fuels. This is very close to an emissions fee but may not be strictly proportionate to GHG emissions.
Levy on Interna-tional Aviation and Maritime Transport	A tax could be levied on activities or per-use of emission-related entities, such as tickets for air travel. This is very close to an emissions fee but may not be strictly proportionate to GHG emissions.
Sovereign Wealth Funds	A publicly owned investment fund, using equity shares, bonds, or other assets (e.g., gold reserves).

Fund Generation Mechanisms	
Special Drawing Rights	Financial reserves held in the International Monetary Fund or a new entity could be issued to public or private participants, possibly in exchange for equity shares in the financial institution, that could be used to raise further capital (e.g., through bond sales) or to disburse as climate financing.
Debt Swap Programs	A country or financial institution holding debt from another country (or conceivably a private entity holding debt) could agree to "swap" that debt (in lieu of repayment) for performing specified actions to mitigate or adapt to climate change, as in Debt- for-Forest Swaps.
Climate Bonds	An entity could issue bonds in order to raise capital for climate-related investments. If not issued by an existing, credit-worthy entity, provisions would need to be made for reserve capital, payment of interest, and other financial requirements.
Foreign Direct Investment	Investment and ownership by entities outside of a country of productive assets, such as low emissions equipment, etc. The foreign investor could acquire shares in an enterprise in exchange for some action (e.g., emission reduction credits), participate in a joint venture, purchase land for forest plantations, etc.
Fund Disbursement Mechanisms	
National Official Development and Climate Change Assistance bilate-ral or multilateral)	Typically bilateral funding as part of overseas development assistance to assist mitigation or adaptation in the context of economic development in low-income countries.
Project-level Emissions Reduction Market	Like other project financing, project developers could seek financing, including concessional financing, in return for getting emission reductions from the project certified and selling them. Initial financing or purchase of the certified emission reductions could be by the private sector, or governments, or some combination.
Program or Sectoral Emissions Reduction Market	Governments or industry associations in a country could sell offsets or certified emission reductions achieved by broad programs (e.g., tighter energy efficiency standards) or sector-wide actions (e.g.,

Table A-1. (Continued)

Fund Generation Mechanisms	
	installation of carbon capture and sequestration on all powerplants).
Reverse Auction	A government or other large entity could request bids and then purchase certified emission reductions offered at the lowest cost per unit (or other criteria). Alternatively, an entity could purchase andaggregate certified emission reductions from a variety of sources and them sell them to highest-bidding private sector or governmental entities.
Grants	A transfer of cash, goods, or services for which no repayment is required. Grants can supplement other forms of financing, including leveraging of private resources.
Performance-based Grants	A transfer of cash, goods, or services for which no repayment is required, but requiring demonstration of performance (i.e., emissions reductions or forest preservation), typically before the entire transfer is made.
Concessional Debt	Transfer of funds (e.g., loans) for which repayment of the funds is required, but at lower-than-market interest rates or other favorable treatment (e.g., extended repayment periods).
Equity	Funding provided in exchange for a share of ownership of a project or entity (i.e., corporation).
Loan Guarantee	A legal commitment by one entity to take on the debt of a borrower if that borrower is unable or unwilling to repay according to the terms of the loan. Loan guarantees could be given for specific projects or for broad program or sectoral investments.

Source: **Appendix A** was prepared by Jane A. Leggett, Specialist in Energy and Environmental Policy, jaleggett@crs.loc.gov, 7-9525. The terms (not definitions) are modified from a list of options in Global Canopy Programme, *The Little Climate Finance Book: A Guide to Financing Options for Forests and Climate Change*, December 2009. http://www.globalcanopy.org; OECD, 2009.

Appendix B. Comparison of Sources of Climate Change Financing

Table B-1. Considerations Concerning Sources of Climate Change Financing

Criteria/ Options	Public Funds, Bilaterally	Public Funds, Multilaterally	GHG Reduction Credit Markets	"Share of Proceeds" of GHG Markets	Other Private Investment	Philanthropic and Other Private
Magnitude likely available[a]	Currently largest portion of pledged fast-start funds	Smaller share than bilateral funding, Perhaps comparable to current GHG credit markets	In long run, potentially the largest trackable quantity, depe-ndent on a policy framework that establishes a premium for GHG reductions (i.e., prices on emission reductions)	Likely a small percentage (e.g., 2%-5%) of the size of GHG markets	Potentially the largest quantity, but distinguishing from non-climate-change investmen may be problematic	Possibly comparable to recent bilateral funding[b]
Predictability	Fair to moderate. Dependent on national Appropriations processes and macro-economic conditions. May be subject to changes in priorities of budgets. Predictability may be improved by legal provisions enacted nationally	Fair to moderate, depending on pledges and prompt payment into multilateral funds. May be subject to changes in priorities of budgets of countries and multi-purpose multilateral funds	Fair to moderate. Dependent on the existence and stability of policy frameworks, energy and macro-economic markets, techn-ological advance, competition among suppliers and purchasers, and other factors. Once legal	To the degree established by rules, the percentage may be highly predict-table. The absolute flow of funds would be dependent on the predictability of the size of GHG reduction credit markets.	Moderate. Dependent on the existence and stability of policy frameworks, energy and macro-economic markets, technological advance, competition among suppliers and purchasers, and other factors. Once legal frameworks are in place, GHG markets may be more predict-table than annual	Likely least predictable of options

Table B-1. (Continued)

Criteria/ Options	Public Funds, Bilaterally	Public Funds, Multilaterally	GHG Reduction Credit Markets	"Share of Proceeds" of GHG Markets	Other Private Investment	Philanthropic and Other Private
	to generate a flow of funds outside of annual or regular appropriations[c]		frame-works are in place, GHG markets may be more predictable than annual governmental appropriations.		governmental appropriations.	
"Additionality"	Difficult to evaluate, in most cases, because of typically variable aid flows, and no projections of baseline aid, The "strength" of mainstream-ming into development priorities makes discerning additionality difficult	Possibly the easiest to track as "additional" to past flows	Difficult to evaluate, Altho-ugh detailed methods and rules have been estab-lished for evalu-ating baselines and projects. Demonstrating additionality makes financing more difficult and slow, and reports exist of inappropriate approvals	As clearly additional as the GHG reductions made	Likely difficult to discern, especially in flows of Foreign Direct Investment and where modifications to investments have been made to mitigate GHG emissions or forest destruction, or to avoid damages	Unknown. However, there may be few incentives to misre-present the additiona-lity of new climate-related initiatives and grants
Access	Typically direct access by countr-ies and other institutions where existing partners-hips and priorities exist in donor	Typically provide broader access than bilateral funding	Dependent on ability of seekers to participate in markets (i.e., sufficient skill, stability, creditw-orthiness, etc.);	Theoretically, open access to any government or entity that is meets eligibility criteria	Access is greatest to private sector projects that would be profi-table without or with further public incen-tives (e.g., a price on carbon or a renewable	Unknown. Access is likely best for larger project ts managed by well established entities with a

Criteria/ Options	Public Funds, Bilaterally	Public Funds, Multilaterally	GHG Reduction Credit Markets	"Share of Proceeds" of GHG Markets	Other Private Investment	Philanthropic and Other Private
	countries. Possibly difficult access for countries that are small and not high priority bilateral partners of donors		Likely access for GHG mitigation, but not adapta-tion. Requeting entity must meet any eligib-ility criteria and present a project that meets standards for transparency and performance.		energy quota). Access is unlikely for Adaptation projects, small projects, and projects without reliable revenue flows, and proponents with poor access to credit	proven record of positive performance. Some philanthropic organizations, however, specialize in higher risk projects and micro-finance. Public and private funds may diminish philanthropic contributions
Transparency	Poor to moderate. Dependent on availability of public reports on details of funding. Poor transpar-ency of performance on funded projects and programs.	High, because of public reporting and Review requirements.	High, because of public reporting and review requirements	Highest because of clear share of proceeds from emission reduction credit markets	Poor to fair, as there is unlikely to be a require-ment for reporting public-ally such flows in detail and with verification	Poor to moderate, as there is unlikely to be a requirement for reporting publically such flows in detail and with verification
Fiduciary Standards	Unclear. Depend-ent on practices of donor and recipient	Moderate and improving.	Theoretically high, but expe-nsive and time-consuming to	Depends on the Requirements of the project review, disbursement, and	Presumably high, as private investors have incentives to set high standards	Varies with the requir-ements of each

Table B-1. (Continued)

Criteria/ Options	Public Funds, Bilaterally	Public Funds, Multilaterally	GHG Reduction Credit Markets	"Share of Proceeds" of GHG Markets	Other Private Investment	Philanthropic and Other Private
	entities, and of willingness to report in detail to public		ensure. Some limited reports of fraud in exis-ting markets. Will depend on effi-cacy of project performance verification as well	accountability mechanism		philanthropic organization
Efficiency	Easiest to "mainstream" Into development priorities	Possibly least efficient the more centralized the review and disbursement mechanism	Financing is directly tied to mitigation perfo-rmance. Theoreti-cally the most efficient, but realistically dependent on the absence of failures or inefficiencies, such as existing externalities, lack of information, unequal access	Possibly least effici-ent the more centra-lized the review and disbursement mechanism, and the greater the requir-ements for project proposal, review, and verification	Theoretically efficient. If Transaction costs rise too high, the investment becomes unprofitable and funds flow to more efficient investment. Efficiency would be compromised if actual project climate-related performance is poor	Varies with the requir-ements of each philan-thropic organization. Likely more efficient than large, public funds with public review and decision-making processes

Source: **Appendix B** was prepared by Jane A. Leggett, Specialist in Energy and Environmental Policy, jaleggett@crs.loc.gov, 7-9525.

a. The judgments in this table about the likely magnitudes of funding are based, in part, on analyses of past flows, pledges, and theoretical analyses of the potentials (based also on historical evidence). See, for example, Jan Corfee-Morlot, Bruno Guay, and Kate M. Larsen, Financing Climate Change Mitigation: Towards a Framework for Measurement, Reporting and Verification

(Paris: Organisation for Economic Cooperation and Development, October 2009), http://www.oecd.org/dataoecd/0/60/44019962.pdf.

b.Although compilations are not available of philanthropic support to address climate change, the magnitude is likely in the billions of dollars, based on press reports. See, for example: http://philanthropy.com/article/ Grant-Makers-Pour-More-Than/56848/; http://philanthropy.com/article/Rockefeller-Commits/62676/;http://philanthropy.com/article/Doris-Duke-Foundation-Gives/54670/; http:// philanthropy.com/article/SorosPledges-100-Million-/57718/; etc.

c.Consideration of mechanisms to assure funding through public institutions has occurred in a number of fora, and has been enacted by the European Commission. In the United States, several legislative proposals (e.g., H.R. 2454, the American Clean Energy and Security Act passed by the House in June 2010) would allocate a portion of revenues generated by the bill to international financing. Whether such revenues would be subject to further appropriation is often controversial. Internationally, a high-level panel convened by United Nations Secretary General Ban Ki-Moon is studying proposals for levies on international bunker fuels, redirection of fossil fuel subsidies, etc. that willing countries might enact to generate a relatively reliable flow of funds.

FY2011 Enacted Budget Authority and the FY2012 Budget Request

Based on provisions in H.R. 1473, most FY2011 enacted funding for international climate programs at the Department of State, the Department of the Treasury, and USAID (Function 150) are drawn from larger line item agency categories. Allocations for these programs are at the discretion of the agency, in relation to its other programming, and have yet to be reported.

APPENDIX C. U.S. GLOBAL CLIMATE CHANGE INITIATIVE

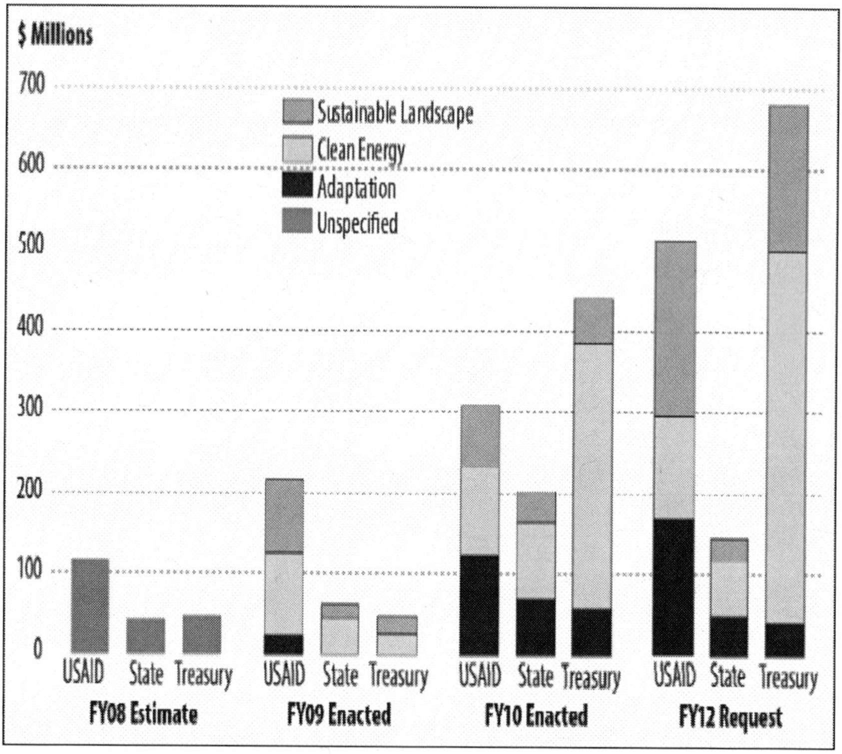

Figure C-1. Recent U.S Budget Authority and Requests for the Obama Administration's Global Climate Change Initiative by Agency and Category.

Table C-1. Recent U.S Budget Authority and Requests for the Obama Administration's Global Climate Change Initiative by Agency and Category. Assistance by Category, Agency, Year, Nominal US$ in Millions

Category	Agency	FY2008 Estimate	FY2009 Enacted	FY2010 Enacted	FY2011 Request	FY2011 Enacted	FY2012 Request
Adaptation	USAID	$15	$24	$123	$187	TBD	$167
	State	0	0	68	57	TBD	49
	Treasury	0	0	55	90	TBD	40
Energy	USAID	50	100	108	129	TBD	129
	State	31	43	94	74	TBD	66
	Treasury	26	26	329	508	TBD	457
Landscapes	USAID	50	90	75	175	TBD	213
	State	10	18	40	25	TBD	28
	Treasury	20	20	54	147	TBD	180
Total		*202*	*321*	*946*	*1,391*	*TBD*	*1,329*

Source: OMB, *Federal Climate Change Expenditures Report to Congress 2007*, 2010; Budget of the USG, 2011, 2012.

Notes: Figures represent annual budget authority for the Department of State, the Department of Treasury, and the U.S. Agency for International Development through appropriations by the U.S. House and Senate Subcommittees on State, Foreign Operations & Related Programs. There were no "Federal Climate Change Expenditures Report to Congress" in 2008 and 2009; the FY2008 figure is an estimate from the 2007 report. Funding was unspecified by category in FY2008; the FY2008 category figures are CRS approximations. For FY20 11, allocations by category have not yet been reported by agencies based on enacted appropriations in H.R. 1473 (P.L. 112-10). Budget authority listed above does not include international programs in complementary agencies (e.g., DOE, EPA, DOC, NSF, USDA, NASA, MCC, TDA).

The Obama Administration's FY2012 budget request for international climate programs at the Department of State, the Department of the Treasury, and USAID (Function 150) would fund near-term climate financing of slightly over $1.3 billion across three areas: (1) clean energy[89] ($652 million, up from $531 million enacted for FY20 10) to reduce net GHG emissions from the energy sector, industry, and urban areas; (2) sustainable landscapes[90] ($421 million, up from $169 million enacted for FY20 10) to assist reduction of GHG emissions from deforestation and land degradation; and (3) adaptation[91] ($256 million, up from $246 million enacted for FY2010) to focus on helping countries manage climate and weather-related risks and build resilience.

"Adaptation" ($256 million FY2012 request, up from $246 million in FY2010)

Assisting countries manage climate and weather-related risks and build climate resilience.

- Bilateral and regional programs: FY2012 req. $167 million, USAID (FY2011 req. $187m; FY2010 en. $123m): U.S. investments that target the more vulnerable countries in Africa, Asia, and Latin America and address climate risks in areas including infrastructure, agriculture, health and water services. They aim to develop capacity for countries to use the best science and analysis for decision making, and promote sound governance to carry out these decisions.

- Least Developed Country Fund, Special Climate Change Fund, and other activities: FY20 12 req. $49 million, State (FY20 1 1 req. $57m; FY20 10 en. $68m): Multilateral funds that focus on climate resilience and food security provisions in the countries with greatest needs.

- Pilot Program for Climate Resilience: FY20 12 req. $40 million, Treasury (FY20 11 req. $90m; FY20 10 en. $55m): A multilateral pilot program that coordinates comprehensive strategies in nine of the poorest and most vulnerable countries to support actions that respond to climate risks.

"Clean Energy" ($652 million FY2012 request, up from $531 million in FY2010)

Mitigating net greenhouse gas emissions from energy sectors using energy efficiency and low-emission technologies.

- Clean Technology Fund: FY20 12 req. $400 million, Treasury (FY 2011 req. $400m; FY2010 en. $300m): Multilateral program that aims to spur large-scale clean energy investments in lower income countries with rapidly growing emissions.
- Bilateral and regional programs: FY2012 req. $129 million, USAID (FY2011 req. $129m; FY2010 en. $108m): U.S. investments for energy sector reform, energy efficiency, clean transport, and low-carbon energy use that support Low Emission Development Strategies and focus on countries that have strong emission reduction potential and a commitment to the principle of low-emissions growth.
- Global Environment Facility: FY2012 req. $37 million, Treasury (FY2011 req. $58m; FY2010 en. $26m) Multilateral financial institution that provides incremental funding for projects that provide global environmental benefits. 26% of GEF's activities promote the demonstration, deployment, and transfer of innovative, low-carbon technologies. Total GEF request across all environmental sectors: FY2012 req. $144 million, Treasury (FY2011 req. $175m; FY2010 en. $87m).
- Program for Scaling-Up Renewable Energy in Low Income Countries: FY20 12 req. $20 million, Treasury (FY20 11 req. $50m; FY20 10 en. $0m). Multilateral program that assists the poorest countries expand energy access and stimulate economic growth through the scaled-up deployment of clean and secure renewable energy strategies.

"Sustainable Landscapes" ($421 million FY2012 request, up from $169 million)

Reducing greenhouse gas emissions from deforestation and land degradation.

- Bilateral and regional programs: FY2012 req. $213 million, USAID (FY2011 req. $175m; FY2010 en. $75m): U.S. investments that support country-driven policies for forest governance, forest cover and land use change monitoring systems, law-based resource management and land tenure, and on-the-ground efforts to halt deforestation and foster sustainable forest-based livelihoods. The program focuses on globally prominent forests in Africa, Asia, and Americas.
- Forest Investment Program: FY20 12 req. $130 million, Treasury (FY2011 req. $95m; FY2010 en. $20m): Multilateral pilot program that addresses the circumstances that lead to deforestation and increased greenhouse gas emissions in select lower income countries. It aims to improve regulation and enforcement, mobilize private financing, and secure the social and economic benefits of sound forest management.
- Global Environment Facility: FY2012 req. $35 million, Treasury (FY2011 req. $32m; FY2010 en. $0m): Multilateral financial institution that provides incremental funding for projects that provide global environmental benefits. Sustainable forest management funding supports projects that protect biodiversity and sustainable land use. Total GEF request across all environmental sectors: FY2012 req. $144 million, Treasury (FY2011 req. $175m; FY2010 en. $87m).

End Notes

[1] Studies include U.S. Global Change Research Program, *Global Climate Change Impacts in the U.S.,* 2009, at http://globalchange.gov/publications/reports/scientific-assessments/us-impacts/key-findings (accessed March 31, 2011); Intergovernmental Panel on Climate Change, *Climate Change 2007: Working Group I: The Physical Basis* (Cambridge, UK: Cambridge University Press, 2007); National Research Council, *Reconciling Observations of Global Temperature Change*, Board on Atmospheric Sciences and Climate (BASC) (Washington, DC: National Academy Press, 2000); National Research Council, *Climate Change Science: An Analysis of Some Key Questions* (Washington, DC: National Academies Press, 2001); National Research Council, *Abrupt Climate Change: Inevitable Surprises* (Washington, DC: National Academies Press, 2002). For background information on climate change science and impacts, see CRS Report RL33 849, *Climate Change: Science and Policy Implications*, by Jane A. Leggett.

[2] In 2009, the CO_2e global average concentration in Earth's atmosphere was about 0.0387% by volume, or 387 ppm.

[3] Carbon Dioxide Equivalent is a metric measure used to compare the emissions from various greenhouse gases based upon their global warming potential (GWP). The greenhouse gases as defined by the UNFCCC include carbon dioxide (CO_2), methane (CH4), nitrous oxide (N2O), and gases (such as chlorofluorocarbons, CFC) that are controlled under the Montreal Protocol to protect the stratospheric ozone layer.

[4] These policies are in conformance with the United Nations Framework Convention on Climate Change (UNFCCC).

[5] This report uses the terms "developing," "developed," "high income," "low income," "emerging," "industrialized," among others, to describe the economic and political development of countries. It should be noted that there is no universally accepted definition or categoryization of these terms. Many commentators note that divisions between these categories in climate-related literature are often arbitrary. Some commentators suggest that global economic development may best be represented as a spectrum of values as opposed to binary categories. Other analyses use the UNFCCC categories of "Annex I," "Annex II," "non-Annex I," and "Least Developed" to refer to the income level of country Parties (see UNFCCC at http://unfccc.int/parties_and_observers/items/2704.php (accessed March 31, 2011)); however, it should be noted that the UNFCCC categories are based on 1992 OECD statistics that have since changed. Annex I countries currently include the United States, the 27 EU member states, Australia, Canada, Iceland, Japan, New Zealand, Norway, Turkey and Russia. All other nations are in the category of non-Annex I. This report will predominantly use the terms "higher income" and "lower income" in an effort to take note of the aforementioned concerns.

[6] For background on the history of international climate change agreements, see CRS Report R40001, *A U.S.-Centric Chronology of the International Climate Change Negotiations*, by Jane A. Leggett.

[7] United Nations Framework Convention on Climate Change, *Full Text of the Convention*, May 9, 1992, U.N.T.S., vol. 1771, No. 30822; S. Treaty Doc No. 102-38. See art. 4, § 3, which states: "Commitments: The developed country Parties and other developed Parties included in Annex II shall provide new and additional financial resources to meet the agreed full costs incurred by developing country Parties in complying with their obligations under Article 12, paragraph 1 (i.e., the communication of information related to implementation including national emissions inventories, national communications of direction, etc.). They shall also provide such financial resources, including for the transfer of technology, needed by the developing country Parties to meet the agreed full incremental costs of implementing measures that are covered by paragraph 1 of this Article (i.e., national development priorities including emission inventories, mitigation and adaptation strategies, promotion of

low-emission development, promotion of sustainable management, scientific cooperation, national communications, etc.) and that are agreed between a developing country Party and the international entity or entities referred to in Article 11, in accordance with that Article (i.e., the financial mechanism of the UNFCCC). The implementation of these commitments shall take into account the need for adequacy and predictability in the flow of funds and the importance of appropriate burden sharing among the developed country Parties," at http ://unfccc. int/essential _background/convention/background/items/1349. php (accessed March 31, 2011).

[8] Which countries provide funds and in what amounts remains an element of negotiation.

[9] See "Copenhagen Accord," United Nations Framework Convention on Climate Change, Report of the Conference of the Parties on its fifteenth session, held in Copenhagen from 7 to 19 December 2009, Decision 2/CP.15, FCCC/CP/2009/1 1/Add. 1, March 30, 2010, § 8, at http://unfccc.int/resource/docs/2009/cop15/eng/1 1a01 .pdf#page=4 (accessed March 31, 2011).

[10] "Americans want to cut foreign aid...to whom?" *The Economist,* April 9, 2010, 15:58 by M.S., at http://www.economist.com/blogs (accessed March 31, 2011).

[11] William Easterly, "The Cartel of Good Intentions," *Foreign Policy*, vol. 131 (July-August 2002), pp. 40-49.

[12] William Easterly and Tobias Pfutze, "Where Does the Money Go? Best and Worst Practices in Foreign Aid," *Journal of Economic Perspectives*, vol. 22, no. 2 (Spring 2008). For more on foreign aid reform, also see CRS Report R40 102, *Foreign Aid Reform: Studies and Recommendations*, by Susan B. Epstein and Matthew C. Weed and CRS Report R40756, *Foreign Aid Reform: Agency Coordination*, by Marian Leonardo Lawson and Susan B. Epstein.

[13] William Easterly, "Can Foreign Aid Buy Growth?," *Journal of Economic Perspectives*, vol. 17, no. 3 (Summer 2003), pp. 23-48.

[14] Carl-Johan Dalgaard and Henrik Hansen, "On Aid, Growth, and Good Policies," *Journal of Development Studies*, vol. 37, no. 6 (August 2001), pp. 17-41.

[15] Craig Burnside and David Dollar, "Aid, Policies, and Growth," *American Economic Review*, vol. 90, no. 4 (September 2000), pp. 847-868.

[16] Statistical studies of the effectiveness of environmental aid are scarce. One performed by the World Bank shows that only 25% of World Bank-financed environmental projects received a "satisfactory" project outcome rating, compared to 100% for education, 86% for health, and 87% for infrastructure (these figures are for the years 2001-2003). For a more complete discussion on the effectiveness of environmental aid, see Robert L. Hicks et al., *Greening Aid?: Understanding the Environmental Impact of Development Assistance*, (New York: Oxford University Press, 2008).

[17] Dambisa Moyo, "Why Foreign Aid is Hurting Africa," *The Wall Street Journal,* March 21, 2009, at http://online.wsj.com/article/SB123758895999200083.html (accessed March 31, 2011).

[18] "Energy Ministers Endorse Clean-Tech Measures, Back CCS Group," *Greenwire*, July 20, 2010. http://www.eenews.net/Greenwire/print/20 10/07/20/4 (accessed March 31, 2011).

[19] See, for example, Robert J. Lempert, Michael E. Schlesinger, and Steve C. Bankes, "When we don't know the costs or the benefits: Adaptive strategies for abating climate change," *Climatic Change* 33, no. 2 (6, 1996): 235-274.

[20] Munich RE Reinsurance, "Overall Picture of Natural Catastrophes in 2010," press release, January 3, 2011, at http://www.munichre.com/en/media (accessed March 31, 2011).

[21] Both the World Bank and U.S. Geological Survey estimate that every dollar spent on disaster preparedness saves seven dollars in disaster response. The World Bank, *Natural Disasters: Counting the Cost,* March 2, 2004, at http://web 0,,contentMDK: 20169861~ menuPK:34458~pagePK:64003015~ piPK:64003012 ~theSitePK:4607,00.html (accessed March 31, 2011).

[22] National Intelligence Council, *National Intelligence Assessment on the National Security Implications of Global Climate Change to 2030, Statement for the Record by Dr. Thomas Fingar, Deputy Director of National Intelligence for Analysis, National Intelligence Council, before the U.S. Congress, House Permanent Select Committee on Intelligence & House Select Committee on Energy Independence and Global Warming,* June 25, 2008, at http://www.dni.gov/nic/ special_climate2030.html (accessed March 31, 2011).

[23] Joint Chiefs of Staff, *JCS Speech: Energy Security Forum,* as delivered by Admiral Mike Mullen, chairman of the Joint Chiefs of Staff, Washington, DC, Wednesday, October 13, 2010, at http://www.jcs.mil/speech (accessed March 31, 2011).

[24] Annex I countries currently include the United States, the 27 EU member states, Australia, Canada, Iceland, Japan, New Zealand, Norway, Turkey and Russia. All other nations are in the category of non-Annex I. Annex II Parties include the industrialized countries that were members of the OECD (Organization for Economic Co-operation and Development) in 1992, but not the countries with economies in transition (the EIT Parties), including the Russian Federation, the Baltic States, and several Central and Eastern European States.

[25] UNFCCC, *Decision 2/CP.15*, op. cit.

[26] This section presents the findings on various cost estimates for mitigation and adaptation needs in lower income countries. The following section, "Methodology," discusses in greater detail the economic assumptions and modeling behind the various estimates.

[27] For a discussion of contributed estimates, see Jan Corfee-Morlot et al., *Financing Climate Change Mitigation: Towards a Framework for Measurement, Reporting, and Verification*, Organisation for Economic Cooperation and Development, October 2009; and World Bank, *Monitoring Climate Finance and ODA*, Issue Brief #1, May 2010.. OECD DAC introduced a system of monitoring aid that targets the objectives of the Rio Conventions (Rio Markers) in 2008 for mitigation and in 2010 for adaptation. The markers identify aid activities that contribute to the objectives of the UNFCCC. Partial data indicate that over the past few years DAC donors have allocated $3 billion to $4 billion per year for mitigation-related aid (about 3%-4% of total ODA). Other aid has been transferred via the UNFCCC Clean Development Mechanism (CDM) as well as through debt instruments at the Multilateral Development Banks (MDB).

[28] Geoengineering technologies, applied to the climate, aim to achieve large-scale and deliberate modifications of the Earth's energy balance in order to reduce temperatures and counteract anthropogenic (i.e., human-made) climate change. For a detailed examination of geoengineering, see CRS Report R41371, *Geoengineering: Governance and Technology Policy*, by Kelsi Bracmort, Richard K. Lattanzio, and Emily C. Barbour.

[29] "Incremental costs" refer to costs above and beyond the costs of a "business-as-usual" strategy. For example, the incremental cost of a proposed wind farm for electricity generation would be the cost of the wind farm above and beyond the cost of a coal-fired generation plant that produced a comparable amount of electricity. For the purposes of this report "incremental costs" and "additional costs" will be used interchangeably.

[30] A gigaton is equivalent to 1 billion tons.

[31] McKinsey & Company, *Pathways to a Low-Carbon Economy, Version 2 of the Global Greenhouse Gas Abatement Cost Curve,* 2009, p. 7. Reading the curve: "the width of each bar represents the potential of that opportunity to reduce GHG emissions in a specific year compared to the business-as- usual development (BAU). The potential of each opportunity assumes aggressive global action starting in 2010 to capture that specific opportunity, and so does not represent a forecast of how each opportunity will develop. The height of each bar represents the average cost of avoiding 1 ton of CO_2e by 2030 through that opportunity. The cost is a weighted average across sub-opportunities, regions, and years. All costs are in 2005 real Euros. The graph is ordered left to right from the lowest-cost abatement opportunities to the highest-cost. The uncertainty can be significant for individual opportunities for both volume and cost estimates, in particular for the Forestry and Agriculture sectors, and for emerging technologies."

[32] "Afforestation" refers to the planting of trees on areas that are not covered with forests. "Reforestation" refers to the re-planting of trees on areas that were once covered with forests.

[33] The abatement figure of 38 $GtCO_2e$ in 2030 is relative to BAU emissions of 70 $GtCO_2e$, which is equivalent to 35% and 40% decrease from 1990 levels, the reference year for the Kyoto Protocol and many current discussions, and sufficient to having a moderate chance of holding global warming below 2 degrees Celsius.

[34] These estimates use constant 2008 US$ from McKinsey & Co. with a dollar-to-Euro exchange rate of $1.25 to €1.00.

[35] Take for example wind or solar power generation. These technologies have significantly higher up-front capital costs compared to more traditional coal- or natural gas-fired generation due to technical and material costs; however, once a wind or solar power generator has been build or installed, the operating costs are significantly lower, due in no small measure to the fact that its fuel source is renewable and does not need to be purchased.

[36] When assessing costs of adaptation activities, terms like "climate-proofing development" and "disaster relief" can take on several different meanings across different studies. Some studies look only at cost to infrastructure development aimed at reducing anticipated damage from future impacts of climate change to a given region, while other studies may look at the costs of climate-proofing infrastructure to the current climate.

[37] Constant 2008 U.S. $ from McKinsey & Co. estimates using dollar-to-Euro exchange rate of $1.25 to €1.00.

[38] "Scaled-up," "new," "additional," "predictable," and "adequate" are all terms stipulated by the UNFCCC. Debate and controversy over the precise definitions of these terms is outlined in this section.

[39] See "Quasi-Private Sector Mechanisms UNFCCC Kyoto Protocol Mechanisms" section for description.

[40] Investigations into potential sources to finance international climate change mitigation and adaptation activities are numerous among economic foundations and environmental organizations. Further, in February 2010, United Nations Secretary General Ban Ki-moon assembled a High Level Advisory Group on Climate Change Financing to identify technically sound and politically feasible potential sources to scale up long-term financing for adaptation, mitigation, technology development and transfer, and capacity building in developing countries. Discussion on potential sources and an assessment on them according to several criteria—including revenue potential, efficiency, equity, incidence, practicality, reliability, additionality and acceptability—can be found at the Group's U.N. website http://www.un.org/wcm/content/site/climatechange/pages/ financeadvisorygroup (accessed March 31, 2011).

[41] "Additionality" means additional to what currently exists or to would otherwise have occurred. Additionality of financing expresses the concern of current aid recipients that donors could merely shift existing development aid into climate-related funds, with no incremental assistance comparable to the extra costs they perceive to be incurring by addressing climate change.

[42] Department of Economic and Social Affairs, *World Economic and Social Survey 2009, Promoting Development, Saving the Planet*, United Nations Publication, 2009.

[43] For examples of surveys or reports on international investment in climate change, see The United Nations Conference on Trade and Development (UNCTAD), *World Investment Report 2010: Investing in a Low-Carbon Economy*, 2010; The World Bank Group's International Finance Corporation's Climate Unit, at http://www.ifc.org/ climatebusiness; or any multinational private corporation's climate change division (e.g., see, among others, General Electric Company, at http://www.ge.com/news/our_viewpoints/energy Ford Motor Company, at http://corporate.ford.com/microsites/ sustainability or Wal-Mart Stores, Inc., at http:// walmartstores. com/Sustainability/ (accessed March 31, 2011).

[44] For a review of climate change issues as related to private investment, see the industry report from Mercer, *Climate Change Scenarios – Implications for Strategic Asset Allocation,* February 15, 2011, at http://www.mercer.com/articles/ 1406410; or any investor service site such as Responsible Investor, at http://www.responsible-investor.com/home/; or any investment private equity firm that concentrates on climate change investments (e.g., among others, Climate Change Capital, at http://www.climatechangecapital. com/home.aspx) (accessed March 31, 2011).

[45] DESA, op. cit.

[46] Anne Hammill et al., *Microfinance and Climate Change Adaptation.* Institute of Development Studies, Sussex, United Kingdom, Bulletin, vol. 39, no. 4 (September 2008). For examples of microfinancing corporations that address climate change, see, among others, Consultative Group to Assist the Poor, at http://www. cgap.org/p/site/c/template.rc/ 1.9.34043/; Green Microfinance, at http://www.greenmicrofinance.org/; The Center for Financial Inclusion, at http://www.centerforfinancialinclusion.org/Page.aspx?pid=1273; Opportunity International, at http://www.opportunity.org/our-work/ (accessed March 31, 2011).

[47] For examples of public-private partnerships for climate change activities, the United Nations has a comprehensive database of voluntary multi-stakeholder initiatives contributing to the implementation of Agenda 21, Rio+5 and the Johannesburg Plan of Implementation (JPOI). It provides a searchable database of U.N. Commission on Sustainable Development (CSD) Partnerships information, publications, events, and links which implement sustainable development goals, at http://www.un.org/esa/dsd/dsd_aofw_par/par_index.shtml (accessed March 31, 2011).

[48] See box "GHG Emissions and Economic Externalities" for the economic assumptions underlying this analysis.

[49] Recent legislative examples of the use of market-based mechanisms to support international climate change financial assistance to lower income countries include the June 2009, House passed H.R. 2454, the American Clean Energy and Security Act (ACESA or Waxman-Markey bill), which had provisions to allow domestic sources to meet their compliance requirements by acquiring up to 1 billion emissions offsets internationally each year, potentially providing a many-billion-dollars stream of private finance for emission abatement projects in developing countries. The bill also would have auctioned a share of domestic allowances to generate funds to help prevent tropical deforestation, build governance and private sector capacities, support cooperation to advance and deploy clean technologies, and to support adaptation to climate change in vulnerable and low-income countries. The parallel bill in the Senate, S. 1733, the Clean Energy Jobs and American Power Act (CEJAPA) or Kerry-Boxer bill, contained similar provisions. Some Members of Congress and advocates had also sought to increase allocation of allowances and/or appropriations for international finance, from $2 billion to $38 billion for international adaptation.

[50] "Climate Change and Development Challenges," seminar series from the Centre for Financial and Management Studies, University of London, at http://www.cefims.ac.uk/cedepapp/124_ web (accessed March 31, 2011).

[51] Some economists view climate change as more than a simple policy problem of marginal abatement costs. See Morgan, M. Granger et al., "Why Conventional Tools for Policy Analysis Are Often Inadequate for Problems of Global Change." *Climatic Change,* vol. 41, no. 3 (March 1, 1999): pp. 271-281, the discussion beginning p. 274.

[52] For further discussion, see CRS Report RL33799, *Climate Change: Design Approaches for a Greenhouse Gas Reduction Program*, by Larry Parker; and CRS Report R41212, *EPA Regulation of Greenhouse Gases: Congressional Responses and Options*, by James E. McCarthy and Larry Parker.

[53] For further explanations of how emission control systems, including cap-and-trade, may work, see CRS Report RL33799, *Climate Change: Design Approaches for a Greenhouse Gas Reduction Program*, by Larry Parker; CRS Report RL34436, *The Role of Offsets in a*

Greenhouse Gas Emissions Cap-and-Trade Program: Potential Benefits and Concerns, by Jonathan L. Ramseur; CRS Report R40242, *Carbon Tax and Greenhouse Gas Control: Options and Considerations for Congress*, by Jonathan L. Ramseur and Larry Parker, and CRS Report R41049, *Climate Change and the EU Emissions Trading Scheme (ETS): Looking to 2020*, by Larry Parker; among other relevant CRS reports on "Climate Change Policy" available at http://www.crs.gov/.

[54] Tradable Permit Systems are currently in place in the United States at the state and regional level. See the Regional Greenhouse Gas Initiative (RGGI) wherein ten Northeastern and Mid-Atlantic states have capped CO_2 emissions from the power sector with the aim of reducing emissions by 10% by 2018 at http://www.rggi.org/home. See also the regional strategy proposed by the Western Climate Initiative at http://www.westernclimateinitiative.org/

[55] The United States is not a Party to the Kyoto Protocol, and has often not supported international efforts to create an international emission trading market, particularly one in which all significant GHG emitting countries (e.g., China, India) are not a Party.

[56] See discussion of the Clean Development Mechanism in the section "Methods for Delivering International Climate Change Financial Assistance."

[57] For more information about the Clean Development Mechanism or the Adaptation Fund, see the UNFCCC website at http://unfccc.int/cooperation

[58] U.S. Office of Management and Budget, *User Charges,* Circular No. A-25 Revised, at http://www. whitehouse.gov/ omb/circulars_a025/ (accessed March 31, 2011).

[59] It should be noted that fees set a price on emissions, not a limit on the quantity; and their primary effect is economic, not environmental. Thus fees may be unrelated to marginal benefits or pollution abatement.

[60] Among higher income countries, fossil fuel revenues are concentrated among five major producers (U.S., U.K. Canada, Australia, Norway). Impacts within countries would be determined by how budgets are adjusted to compensate for this diversion of existing revenues.

[61] "Concessional loans" are loans that are extended on terms substantially more generous than market loans. The concessionality is achieved either through interest rates below those available on the market or by long grace periods, or a combination of these. See International Monetary Fund, *External Debt Statistics: Guide for Compilers and Users – Appendix III, Glossary,* 2003.

[62] These arguments are further outlined in United Nations, *Report of the Secretary-General's High Level Advisory Group on Climate Change Financing,* November 5, 2010, at http://www.un.org/wcm/webdav/site/climatechange/shared/Documents/AGF_reports/AGF %20Report.pdf (accessed March 31, 2011); and responses to it, such as Timme van Melle et al., *International Climate Financing: From Cancún to a 2°C stabilization pathway*, Ecofys Netherlands BV, 2011, at http://www.fao.org/fileadmin/user_upload/ rome2007/ docs/ Climate_financing_after_Cancun%20ECOFYS.pdf (accessed March 31, 2011).

[63] Proposed to the UNFCCC by the Group of 77 and China. The Group of 77, see http://www.g77 .org/doc/.

[64] See DESA and UN, *Report of the Secretary-General's High Level Advisory Group on Climate Change*, op. cit.

[65] As outlined in the most recent UNFCCC draft negotiations: United Nations Convention on Climate Change, *Draft decision -/CP. 16, Outcome of the work of the Ad Hoc Working Group on long-term Cooperative Action under the Convention,* 2010, Section IV.A. "Finance," at http://unfccc.int/files/meetings/cop_16/application/ pdf/cop16_lca.pdf (accessed March 31, 2011).

[66] In general, "bilateral" assistance involves a direct transfer from one country to another; "multilateral" assistance is distributed through international organizations and agencies like the United Nations Organization and the World Bank Group. Further clarifications and debates on the mechanisms respective merits are elaborated in this section.

[67] Strong arguments exist on both sides as to whether "globalization" is a benefit to developing countries, the environment, or both.

[68] World Bank, 2010, op. cit.

[69] See CRS Report 98-568, *Export-Import Bank: Background and Legislative Issues*, by Shayerah Ilias.

[70] See CRS Report R41170, *Multilateral Development Banks: Overview and Issues for Congress*, by Rebecca M. Nelson.

[71] Danieli Violetti, "Clean Development Mechanism: Achievements and Developments," paper presented at the 6th Session of the High-Level Task Force on the Implementation of the Right to Development, Geneva, 2010.

[72] See CRS Report RL33826, *Climate Change: The Kyoto Protocol, Bali "Action Plan," and International Actions*, by Jane A. Leggett.

[73] For more on the choice between bilateral and multilateral aid, see, for example: Helen Milner and Dustin Tingley, "The Choice for Multilateralism: Foreign Aid and American Foreign Policy," Working Paper, February 10, 2010, and Helen Milner, "Why Multilateralism? Foreign Aid and Domestic Principal-Agent Problems," in *Delegation and Agency in International Organizations*, eds. Darren Hawkins et al. (New York: Cambridge UP, 2006), pp. 107-13 9.

[74] For a broader and more detailed discussion of ODA, please see CRS Report R40213, *Foreign Aid: An Introduction to U.S. Programs and Policy*, by Curt Tarnoff and Marian Leonardo Lawson.

[75] Jan Corfee-Morlot et al., *Financing Climate Change Mitigation: Towards a Framework for Measurement, Reporting, and Verification*, Organization for Economic Cooperation and Development, October 2009. Table 1.

[76] See CRS Report R41 170, *Multilateral Development Banks: Overview and Issues for Congress*, by Rebecca M. Nelson.

[77] See CRS Report RL3 1286, *Debt-for-Nature Initiatives and the Tropical Forest Conservation Act: Status and Implementation*, by Pervaze A. Sheikh.

[78] See CRS Report R41 165, *Global Environment Facility (GEF): An Overview*, by Richard K. Lattanzio.

[79] See CRS Report R41302, *Climate Investment Funds (CIFs): An Overview*, by Richard K. Lattanzio.

[80] See CRS Report R41302, *Climate Investment Funds (CIFs): An Overview*, by Richard K. Lattanzio.

[81] See The UNFCCC "Cancun Agreement" at United Nations Convention on Climate Change, *Draft decision -/CP. 16, Outcome of the work of the Ad Hoc Working Group on long-term Cooperative Action under the Convention,* 2010, Section IV.A. "Finance," at http://unfccc. int/files/meetings/cop_16/application/pdf/ cop16 _lca.pdf.

[82] For more substantive analysis of foreign aid and congressional roles, see CRS Report R40213, *Foreign Aid: An Introduction to U.S. Programs and Policy*, by Curt Tarnoff and Marian Leonardo Lawson; and CRS Report R41 170, *Multilateral Development Banks: Overview and Issues for Congress*, by Rebecca M. Nelson.

[83] A much smaller percentage of funding for international climate change initiatives is included in the program budgets of "complementary agencies" in the U.S. government, including the Department of Energy, the Environmental Protection Agency, the Department of Commerce, the National Science Foundation, the Department of Agriculture, the National Aeronautics and Space Administration, the Millennium Challenge Corporation, and the Trade and Development Agency. See Table 4, Note.

[84] United Nations Conference on the Human Environment, *Declaration of the United Nations Conference on the Human Environment,* June 16, 1972, Principle 12, "Resources should be made available to preserve and improve the environment, taking into account the circumstances and particular requirements of developing countries and any costs which may emanate- from their incorporating environmental safeguards into their development

planning and the need for making available to them, upon their request, additional international technical and financial assistance for this purpose." UNCHE refers to the additional cost principal as "additionality." This report uses "incrementality" so as not to confuse with the prior uses of the term "additionality" in the report.

[85] *Agenda 21* has 40 chapters and 115 program areas laid out over 800 pages, divided into four broad areas, including "Means of Implementation," wherein it addresses how international and national support should be organized, including the transfer to lower income countries of financial resources and environmentally-friendly technology; building capacity through technical assistance, environmental education, and scientific information; creating better databases to bridge the data gaps between nations; and improving international environmental organizations, coordination, and legal processes. United Nations Conference on Environment and Development, *Agenda 21,* 1992, at http://www.un.org/esa/dsd/agenda 21/ (accessed March 31, 2011).

[86] James Gustave Speth and Peter Haas, *Global Environmental Governance* (Washington: Island Press, 2006), p.72.

[87] For more information on international negotiations regarding climate change, see CRS Report R40001, *A U.S.- Centric Chronology of the International Climate Change Negotiations*, by Jane A. Leggett.

[88] Although the United States declines to consider a defined percentage as an appropriate means to share the pledged financing, past responsibilities in similar international fora suggest to some that the United States might provide between 20%-30% of the amount, or $6 billion-$10 billion over three years.

[89] "Clean Energy" programming defined as mitigating net greenhouse gas emissions from energy sectors using energy efficiency and low-emission technologies.

[90] "Sustainable Landscapes" programming defined as reducing greenhouse gas emissions from deforestation and land degradation.

[91] "Adaptation" programming is defined as assisting countries manage climate and weather-related risks and build climate resilience.

In: International Financing Efforts ...
Editor: Randy N. Matthews
ISBN: 978-1-62100-702-9

Chapter 2

THE GLOBAL CLIMATE CHANGE INITIATIVE (GCCI): BUDGET AUTHORITY AND REQUEST, FY2008-FY2012

Richard K. Lattanzio

SUMMARY

The United States supports international financial assistance for global climate change initiatives in developing countries. Under the Obama Administration, this assistance has been articulated primarily as the Global Climate Change Initiative (GCCI), a major strategic platform within the President's 2010 Policy Directive on Global Development. The GCCI aims to integrate climate change considerations into relevant foreign assistance through a full range of bilateral, multilateral, and private sector mechanisms to foster low-carbon growth, promote sustainable and resilient societies, and reduce emissions from deforestation and land degradation. The GCCI is implemented through programs at the Department of State, the Department of the Treasury, and the U.S. Agency for International Development (USAID), and is funded through the Administration's Executive Budget, Function 150 account, for State, Foreign Operations, and Related Programs.

Congress is responsible for several activities in regard to the GCCI, including (1) authorizing periodic appropriations for federal agency programs and multilateral fund contributions, (2) enacting those appropriations, (3) providing guidance to the agencies, and (4) overseeing U.S. interests in the programs and the multilateral funds. Recent budget authority for the GCCI was $323 million in FY2009 and $1,003 million in FY2010, and has been enacted through legislation including the Omnibus Appropriations Act, 2009 (H.R. 1105; P.L. 111-8); the Consolidated Appropriations Act, 2010 (H.R. 3288/div.F; P.L. 111-117); the Supplemental Appropriations Act, 2010 (H.R. 4899; P.L. 111-212); and the Department of Defense and Full-Year Continuing Appropriations Act, 2011 (H.R. 1473; P.L. 112-10). The Administration's GCCI FY2012 budget request is $1,328 million. Congressional committees of jurisdiction for the GCCI include the U.S. House of Representatives Committees on Foreign Affairs (various subcommittees); Financial Services, Subcommittee on International Monetary Policy and Trade; and Appropriations, Subcommittee on State, Foreign Operations, and Related Programs; and the U.S. Senate Committees on Foreign Relations, Subcommittee on International Development and Foreign Assistance, Economic Affairs, and International Environmental Protection; and Appropriations, Subcommittee on State, Foreign Operations, and Related Programs.

As Congress considers potential authorizations and/or appropriations for initiatives administered through the GCCI, it may have questions concerning U.S. agency initiatives and current bilateral and multilateral programs that address global climate change. Some potential concerns may include cost, purpose, direction, efficiency, and effectiveness, as well as the GCCI's relationship to industry, investment, humanitarian efforts, national security, and international leadership. This report serves as a brief overview of the GCCI, its structure, intents, and funding history.

THE GLOBAL CLIMATE CHANGE INITIATIVE

On September 22, 2010, President Obama signed the Presidential Policy Directive on Global Development.[1] The directive called for the elevation of foreign development assistance as a national priority and outlined an integrated approach to development, diplomacy, and national security. The Global Climate Change Initiative (GCCI)—one of the three main pillars to the 2010 directive[2]—aims to integrate climate change considerations into relevant foreign assistance through the full range of bilateral, multilateral, and private mechanisms to foster low-carbon growth, promote sustainable and resilient societies, and reduce emissions from deforestation and land degradation.

The GCCI is divided into three main programmatic initiatives, or categories: (1) clean energy assistance, (2) sustainable landscapes assistance, and (3) adaptation assistance.

Clean Energy

Clean energy programs aim to reduce greenhouse gas emissions from energy generation and energy use by accelerating the deployment of clean energy technologies, policies, and practices. The United States delivers much of its assistance for clean energy deployment through multilateral trust funds. These funds are primarily housed in international financial institutions (e.g., the World Bank), are currently supported by the financial contributions of donor country governments, and provide financial assistance for projects implemented by a variety of organizations, including U.N. agencies, multilateral development banks, nongovernmental organizations, and national institutions. These funds take advantage of existing large-scale greenhouse gas reduction opportunities and establish investment channels for larger private sector financing. They include the Clean Technology Fund,[3] which aims to spur large-scale clean energy investments in lower-income countries with rapidly growing emissions; the Global Environment Facility,[4] which provides incremental funding[5] for energy and infrastructure projects that support global environmental benefits; and the Program for Scaling-Up Renewable Energy in Low Income Countries,[6] which endeavors to assist the poorest countries expand energy access and stimulate economic growth through the scaled-up deployment of renewable energy strategies. Bilateral efforts at the Department of State and U.S. Agency for International Development (USAID) seek to complement the multilateral investments by helping to shape development policy and regulatory environments in the recipient countries.

Sustainable Landscapes

Sustainable landscape programs aim to reduce greenhouse gas emissions from deforestation and forest degradation. Bilateral and regional programs at the Department of State and USAID support country-driven policies for forest governance, forest cover and land use change monitoring systems, law-based resource management and land tenure, and on-the-ground efforts to halt deforestation and foster sustainable forest-based livelihoods. Multilateral initiatives include the Forest Investment Program,[7] which tries to address the circumstances that lead to deforestation and increased greenhouse gas emissions in select lower-income countries by improving regulation and enforcement, mobilizing private financing, and securing the social and economic benefits of sound forest management; and the Global Environment Facility, which provides incremental funding for projects that support global environmental benefits such as biodiversity and sustainable land use.

Climate Change Resilience and Adaptation

Adaptation programs aim to assist low-income countries reduce their vulnerability to climate change impacts and build climate resilience. Bilateral and regional programs at the Department of State and USAID target the more vulnerable countries in Africa, Asia, and Latin America and strive to address climate risks in areas including infrastructure, agriculture, health, and water services; to develop capacity for countries to use the best science and analysis for decision making; and to promote sound governance to carry out these decisions. Multilateral initiatives supported by the United States include the Least Developed Country Fund[8] and the Special Climate Change Fund,[9] which focus on climate resilience and food security provisions in countries with the greatest needs; and the Pilot Program for Climate Resilience,[10] which is tasked with coordinating comprehensive strategies in nine of the most vulnerable countries to support actions that respond to the potential risks of a changing climate.

BUDGET AUTHORITY

The Global Climate Change Initiative is funded through programs at the Department of State, the Department of the Treasury, and USAID (i.e., GCCI "core" agencies). Funds for these programs are appropriated in the Administration's Executive Budget, Function 150 account, for State, Foreign Operations, and Related Programs. Budget authority for GCCI programming has increased approximately 500% since FY2008, and currently accounts for approximately 2% of the total State, Foreign Operations, and Related Programs budget (approximately 3% if considering only the Foreign Operations budget).[11] Recent budget authority for the GCCI was a reported \$323 million in FY2009 and \$1,003 million in FY2010.[12] Allocations for

GCCI programming from the enacted FY2011 appropriations are still under agency consideration. The Administration's FY2012 GCCI budget request was $1,328 million. Some additional funds for international climate change financing flow through programs at complementary agencies within the federal government (e.g., the Department of Energy, the Environmental Protection Agency, the Department of Agriculture); however, these allocations are defined outside of the GCCI. Budget authority for GCCI programming in core agencies from FY2008 to FY2012 is represented in Figure 1. Funding levels by category and by agency account, in both core and complementary agencies, from FY2008 to FY2012 are presented in Table A-1 and Table A-2, respectively.

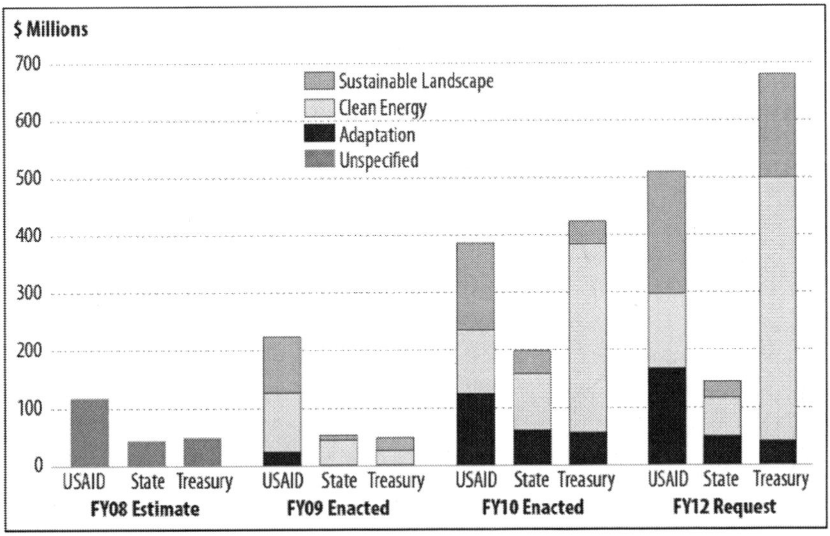

Source: U.S. Department of State, Executive Budget Summary: Function 150 and Other International Programs, Fiscal Year 2012; U.S. Department of State, FY2011 Budget for International Climate Change Financing: An Interagency Strategy Document. Office of Management and Budget, Federal Climate Change Expenditure Reports to Congress, May 2007 and June 2010.

Notes: Enacted funding for the GCCI in FY2008 was not classified by category (see **Table A-2** for FY2008 figures by agency). FY2011 is not represented in the figure because funding for FY2011, enacted under P.L. 112-10, is at the account level, and allocations for many GCCI sub-account programmatic activities have yet to be reported by the agencies (see **Table A-2**).

Figure 1. Global Climate Change Initiative: Budget Authority FY2008-FY2012 (US$ in millions, nominal).

Congress is responsible for several activities in regard to the GCCI, including (1) authorizing periodic appropriations for federal agency programs and multilateral fund contributions, (2) enacting those appropriations, (3) providing guidance to the agencies, and (4) overseeing U.S. interests in the programs.

Congressional committees of jurisdiction for international climate change programs at the Department of State, the Department of the Treasury, and USAID include the following:

- the U.S. House of Representatives Committee on Foreign Affairs (various subcommittees)
- the U.S. House of Representatives Committee on Financial Services, Subcommittee on International Monetary Policy and Trade
- the U.S. House of Representatives Committee on Appropriations, Subcommittee on State, Foreign Operations, and Related Programs
- the U.S. Senate Committee on Foreign Relations, Subcommittee on International Development and Foreign Assistance, Economic Affairs, and International Environmental Protection
- the U.S. Senate Committee on Appropriations, Subcommittee on State, Foreign Operations, and Related Programs

FY2011 Budget Authority

H.R. 1473, the Department of Defense and Full-Year Continuing Appropriations Act, 2011, was enacted April 15, 2011, as P.L. 112-10. It included appropriations for the Department of State, Foreign Operations, and Related Programs (Title IX) that support Global Climate Change Initiative programming. Many GCCI activities are funded by allocations at the sub-account level and were left undefined in P.L. 112-10. Thus, allocations for FY2011 GCCI sub-account programmatic activities are at the discretion of the agencies, under committee consultation, and have yet to be fully reported.

Appropriations enacted in P.L. 112-10 at the account level related to GCCI activities include the following:[13]

- $90 million for the Global Environment Facility (of which the Global Environment Facility reports that approximately 51% is allocated toward projects related to global climate change activities, with the remainder allocated to projects related to biodiversity, international waters, ozone protection, organic pollutants, etc.)

- $185 million for the Clean Technology Fund
- $50 million for the Strategic Climate Fund (including the Pilot Program for Climate Resilience, the Forest Investment Program, and the Program for Scaling-Up Renewable Energy in Low Income Countries)
- Larger account level appropriations at the Department of State, USAID, and the Department of the Treasury from which are drawn funds for sub-account level programming for GCCI activities. These larger accounts include "Development Assistance" at $2,525 million; "Assistance for Europe, Eurasia and Central Asia" at $697 million; "Economic Support Fund" at $5,958 million; "Department of the Treasury, Debt Restructuring" at $50 million; and "International Organizations and Programs" at $355 million, of which up to $10 million was made available for the International Panel on Climate Change/United Nations Framework Convention on Climate Change

See the **Appendix** for a breakdown of the FY2011 request and enacted budget authority.

FY2012 Budget Request

The President's FY2012 budget request follows on the December 2010 United Nations Framework Convention on Climate Change (UNFCCC) negotiations in Cancun, Mexico,[14] which formulated a package of "nationally appropriate"[15] measures toward the goal of avoiding dangerous climate change. These measures include mitigation actions, verification systems, an adaptation framework, a forestry program, a funding mechanism, and a commitment to near-term and long-term climate financing for the least developed countries amounting to near $30 billion for the period 2010-2012 and $100 billion annually by 2020. The President's FY2012 budget requests for international climate programs at the Department of State, the Department of the Treasury, and USAID would fund near-term climate financing of slightly over $1.3 billion across three areas:

- adaptation ($255 million, up from $238 million enacted for FY2010)
- clean energy ($652 million, up from $531 million enacted for FY2010)
- sustainable landscapes ($421 million, up from $232 million enacted for FY2010)

See the Appendix for a breakdown of the FY2012 request.

KEY ISSUES FOR CONGRESS

As Congress considers potential authorizations and/or appropriations for initiatives administered through the Department of State, the Department of the Treasury, USAID, and other agencies with international programs, it may have questions concerning the purpose, direction, efficiency, and effectiveness of U.S. agency initiatives and current bilateral and multilateral programs that address global climate change.

Some issues that may raise concerns over providing assistance include:

- Fiscal Constraints. Budget constraints may lead to questions about sustaining high levels of support for international development assistance in general, and international climate change assistance in particular. The burden is exacerbated during times of economic downturn, when the federal government is hard-pressed to generate fiscal resources to adequately address domestic challenges and maintain basic levels of public services and quality of life. Some have suggested that retaining available funds for immediate domestic priorities, such as fostering renewed economic growth and creating jobs, should take precedence over global concerns for which many Americans feel less urgency and responsibility.

- Potential for Misuse. National and international institutions that dispense financial assistance have sometimes been criticized for inefficient and bloated bureaucracies, their lack of transparency about project procurement practices and operating costs, and the proportion of their funds misused or lost through instances of graft, corruption, and other political inefficiencies.[16] Some suggest revisiting operational guidance of these institutions before further appropriations are made.

- Uncertain Results. Questions remain regarding the overall effectiveness of international financial assistance in spurring economic development and reform in lower-income countries, and, more specifically, in addressing issues of climate change and the environment. Many studies have examined the effects of international assistance provided to lower-income countries, including both bilateral and multilateral mechanisms, and have returned mixed results, making it difficult to reach firm conclusions and support continued contributions.

- Lack of Consensus on Climate Science. Current uncertainties and ambiguities regarding the fields of atmospheric chemistry and climatology have been offered by some as reasons to postpone and/or reconsider international climate change assistance policies and programs.

Some issues that may support providing assistance include:

- Commercial Interests. Some maintain that international climate change assistance benefits U.S. businesses, as support for low-emission economic growth may increase trade, commerce, and economic activity in the global marketplace for domestic goods and services. Increased U.S. funding may allow some U.S. industries to make competitive inroads into rapidly expanding markets, improve the advancement and commercialization of U.S. technologies, mobilize greater investment in related domestic sectors, and enhance job creation in the United States. Decreased funding may cede American influence in global markets to other economic powers still engaged with lower-income countries on environmental and natural resource issues (e.g., the European Union, China).
- Investment Efficiencies. Some argue that the costs of responding to tomorrow's climate-related catastrophes, instabilities, conflicts, and technological needs may be significantly higher than the costs of working today to prevent them through emissions reductions. Some economists note that lower-income countries account for nearly all of the recent growth in global emissions and represent the cheapest near-term opportunity to mitigate GHG pollution as part of a cost-effective solution.

- Natural Disaster Preparedness. Some claim that international climate change assistance is a means to support natural disaster preparedness around the globe. Assistance for adaptation activities to help "climate-proof" developing countries' infrastructure and other sectors may help avoid capital and other losses, minimize the redirection of resources to ad hoc disaster response and urgent humanitarian needs, and avoid chronic humanitarian crises, such as food shortages, particularly for the resource poor in the least developed countries.[17]

- National Security. Some defend international climate change assistance as a way to address and mitigate risks to national security. According to a 2008 National Intelligence Assessment, the impacts of global climate change may worsen problems of poverty, social tensions, environmental degradation, and weak political institutions across the developing world.[18] Some see international climate change assistance as a means to help make lower-income countries less susceptible to these threats, for the benefit of both the country and the security interests of the United States.

- International Leadership. Finally, some see the promotion of international climate change assistance to lower-income countries as a method through which to increase U.S. leadership in global environmental issues. Through such leadership, the United States may be able to influence and set important international economic and environmental policies, practices, and standards.[19]

APPENDIX. GCCI BUDGET TABLES

Table A-1. Global Climate Change Initiative, Core Agencies Budget Authority FY2009-FY2012, by Category (US$ in millions; figures appearing as an asterisk—*—have yet to be reported by the agencies).

Category/Core Agency[a]	FY2009 Actual	FY2010 Enacted	FY2011 Request	FY2011 Enacted	FY2012 Request
Adaptation[b]	**24**	**238**	**334**	*	**255**
State	0	60	57	*	49
Treasury	0	55	90	*	40

USAID	24	123	187	*	166
Clean Energy[c]	**169**	**531**	**711**	*	**652**
State	43	96	74	*	66
Treasury	26	326	508	*	457
USAID	100	109	129	*	129
Sustainable Landscapes[d]	**128**	**232**	**346**	*	**421**
State	10	40	25	*	28
Treasury	20	40	146	*	180
USAID	98	152	175	*	213
Total State	**53**	**196**	**156**	*	**143**
Total Treasury	**46**	**421**	**744**	*	**677**
Total USAID	**222**	**384**	**491**	*	**508**
Total	**321**	**1,001**	**1,391**	*	**1,328**

Source: U.S. Department of State, Congressional Budget Justification, Volume 2, Foreign Operations, Fiscal Year 2012; Office of Management and Budget, Federal Climate Change Expenditure Reports to Congress, June 2010; and U.S. Department of State, FY2011 Budget for International Climate Change Financing: An Interagency Strategy Document.

Notes: Budget authority provided in millions of dollars. Discrepancies may result from rounding and improved estimates. Mild discrepancies exist between Table A-1 and Table A-2 due to the existence of different source material for funding reported by category versus funding reported by agency account for FY2008-FY2010. Figures appearing as an asterisk—*—represent funding for programs at the sub-account level for which allocations are drawn from larger account level programming and have yet to be reported by the agencies. Funding for the GCCI in FY2008 was not classified by category, and is not included in this table (see **Table A-2** for FY2008 figures by agency).

a. GCCI "core agencies" include the Department of State, the Department of the Treasury, and USAID as represented in the Department of State's Executive Budget, Function 150 and Other International Programs (see **Table A-2** for a summary of GCCI core and complementary agencies' budget allocations by account).

b. Adaptation programs aim to assist low-income countries reduce their vulnerability to climate change impacts and build climate resilience.

c. Clean Energy programs aim to reduce greenhouse gas emissions from energy generation and energy use by accelerating the deployment of clean energy technologies, policies, and practices.

d. Sustainable Landscape programs aim to reduce greenhouse gas emissions from deforestation and forest degradation.

Table A-2. Global Climate Change Initiative, All Agencies Budget Authority FY2008-FY2012, by Program (US$ in millions; figures appearing as an asterisk—*—have yet to be reported by the agencies).

Agency/Account	FY2008 Estimate[a]	FY2009 Actual[b]	FY2010 Enacted[c]	FY2011 Request[d]	FY2011 Enacted[e]	FY2012 Request[f]
Department of State[g]	**41**	**55**	**199**	**156**	*	**142**
Diplomatic and Consular Affairs[h]	4 [6,789]	2 [7,153]	2 [12,599]	—	—	—
Economic Support Fund[i]	32 [5,323]	24 [7,105]	158 [8,834]	110 [7,812]	* [5,958]	99 [7,185]
International Organizations and Programs[j]	5 [317]	29 [352]	39 [394]	46 [351]	* [355]	43 [349]
Department of the Treasury[k]	**46**	**46**	**421**	**744**	*	**677**
Debt Restructuring: Tropical Forest Conservation Act[l]	20 [30]	20 [60]	20 [60]	20 [70]	* [50]	15 [15]
Global Environment Facility[m]	26 [81]	26 [80]	26 [87]	89 [175]	46 [90]	72 [144]
Clean Technology Fund[n]	0	0	300	400	185	400
Strategic Climate Fund: Pilot Program for Climate Resilience[o]	0	0	55	90	* [50]	40
Strategic Climate Fund: Forest Investment Program[p]	0	0	20	95	* [50]	130
Strategic Climate Fund: Scaling Up Renewable Energy Program[q]	0	0	0	50	* [50]	20
U.S. Agency for International Development[r]	**115**	**222**	**383**	**491**	*	**509**
Assistance for Europe, Eurasia, and Central Asia[s]	26 [690]	15 [922]	26 [742]	26 [716]	* [697]	29 [627]
Development Assistance[t]	81 [1,624]	113 [2,000]	313 [2,520]	446 [2,981]	* [2,525]	452 [2,918]

Agency/Account	FY2008 Estimate[a]	FY2009 Actual[b]	FY2010 Enacted[c]	FY2011 Request[d]	FY2011 Enacted[e]	FY2012 Request[f]
International Disaster Assistance[u]	2 [650]	—	—	—	—	—
Economic Support Fund[v]	7 [5,323]	94 [7,105]	44 [8,834]	19 [7,812]	* [5,958]	28 [7,185]
Total (Core Agencies)	**202**	**323**	**1,003**	**1,391**	*	**1,328**
Complementary Agencies[w]	*	50	77	110	*	*
Total (All Agencies)	**202**	**373**	**1,080**	**1,501**	*	*

Source: See notes a-f below for sources for each fiscal year.

Notes: Budget authority provided in millions of dollars. Discrepancies may result from rounding and improved estimates. Mild discrepancies exist between **Table A-1** and **Table A-2** due to the existence of different source material for funding reported by "category" versus funding reported by "agency account."

Figures appearing in brackets—[ ]—represent total budget authority for all activities within the account, from which sub-account allocations for GCCI programs are represented by the adjoining, non-bracketed figure.

Figures appearing as an asterisk—*—represent funding for programs at the sub-account level for which allocations have yet to be reported by the agencies.

a. Office of Management and Budget, Federal Climate Change Expenditure Reports to Congress, May 2007. These figures are estimates based upon the Administration's request. There were no Expenditure Reports published for FY2009 or FY2010 to report actual or enacted funding levels for prior years, and thus sub-account allocations for GCCI programming during these years remain unconfirmed. Account level budget authority from FY2008 Cons. Enacted (P.L. 110-161); FY2008 Supp. Enacted (P.L. 110-252).

b. Office of Management and Budget, Federal Climate Change Expenditure Reports to Congress, June 2010; account level budget authority from FY2009 Base Enacted (P.L. 111-8); FY2009 Bridge Enacted (P.L. 110-252); FY2009 Stimulus Enacted (P.L. 111-5); FY2009 Supp. Enacted (P.L. 111-32).

c. Office of Management and Budget, Federal Climate Change Expenditure Reports to Congress, June 2010; account level budget authority from FY2010 Base Enacted (H.R. 3288/div.F; P.L. 111-117); FY2010 Supp. Enacted (H.R. 4899; P.L. 111-212).

d. U.S. Department of State, Congressional Budget Justification, Volume 2, Foreign Operations, FY2011.

e. FY2011 Continuing Resolution Enacted (H.R. 1473; P.L. 112-10). The figures in this table do not reflect the 0.2% rescission across all non-defense accounts for FY2011 funds, in accordance with Section 1119(a) of P.L. 112-10.

f. U.S. Department of State, Congressional Budget Justification, Volume 2, Foreign Operations, FY2012.

g. U.S. Department of State GCCI figures not reported as actual are considered estimates and may change as offices, bureaus, and/or missions update operational budgets and outlays are finalized.

h. Diplomatic and Consular Affairs includes sub-account programmatic activities to support international climate change activities; but because it is not a foreign assistance account, it has been excluded from the agency cross-cut beginning in FY2011. This funding was not included in the agencies' category allocations as reported for FY2011, and thus does not appear in **Table A-1**.

i. Economic Support Fund includes sub-account programmatic activities to support international climate change activities, including funding for the Office of Oceans and International Environmental and Scientific Affairs (which contributes funds for adaptation programming at the U.N. Least Developed Countries Fund (LDCF) and the U.N. Special Climate Change Fund (SCCF), clean energy programming at the Major Economies Forum/Clean Energy Ministerial and Climate Renewables and Efficiency Deployment Initiative, and the Global Methane Initiative, and sustainable landscape programming at the Forest Carbon Partnership Facility), and the Office of Western Hemisphere (which contributes to the Energy and Climate Partnership of the Americas).

j. International Organizations and Programs includes sub-account programmatic activities to support international climate change activities, including contributions to the International Panel on Climate Change, UN Framework Convention on Climate Change, and the Montreal Protocol Multilateral Fund. For FY2011, P.L. 112-10 enacted $355 million to the account, of which up to $10 million may be made available for the International Panel on Climate Change/United Nations Framework Convention on Climate Change.

k. U.S. Department of the Treasury GCCI figures not reported as actual are considered estimates and may change as offices update operational budgets and outlays are finalized.

l. Debt Restructuring includes sub-account programmatic activities to support international climate change activities in sustainable landscapes programming including funding for the Tropical Forest Conservation Act.

m. Global Environment Facility (GEF) is a multilateral environmental trust fund that supports climate change activities in clean energy and sustainable landscapes programming. Established in 1991, the United States has contributed funds annually since 1993. Only a portion of GEF funds—as determined through GEF programming decisions—is allocated to climate change activities; the remaining allocation supports other environmental sectors (e.g., biodiversity, oceans, ozone, chemicals).

n. Clean Technology Fund is a multilateral environmental trust fund that supports climate change activities in clean energy programming. Established in 2008, the United States has contributed funds annually since 2010.

o. Strategic Climate Fund: Pilot Program for Climate Resilience is a multilateral environmental trust fund that supports climate change activities in adaptation programming. Established in 2008, the United States has contributed funds annually since 2010. FY2011 Continuing Resolution (H.R. 1473; P.L. 112-10) enacted budget authority of $50 million for the Strategic Climate Fund; allocations between the three sub-programs are at the discretion of the Department of the Treasury and have yet to be reported.

p. Strategic Climate Fund: Forest Investment Program is a multilateral environmental trust fund that supports climate change activities in sustainable landscapes programming. Established in 2008, the United States has contributed funds annually since 2010. FY2011 Continuing Resolution (H.R. 1473; P.L. 112-10) enacted budget authority of $50 million for the Strategic Climate Fund; allocations between the three sub-programs are at the discretion of the Department of the Treasury and have yet to be reported.

q. Strategic Climate Fund: Scaling Up Renewable Energy Program is a multilateral environmental trust fund that supports climate change activities in clean energy programming. Established in 2008, the United States did not contribute funds in 2010 because the program was not yet operational. FY2011 Continuing Resolution (H.R. 1473; P.L. 112-10) enacted budget authority of $50 million for the Strategic Climate Fund; allocations between the three sub-programs are at the discretion of the Department of the Treasury and have yet to be reported.

r. U.S. Agency for International Development (USAID) GCCI figures not reported as actual are considered estimates and may change as offices update operational budgets and outlays are finalized.

s. Assistance for Europe, Eurasia, and Central Asia includes sub-account programmatic activities to support USAID national, regional, and central international climate change activities. Prior to FY2009, this account was two separate accounts: Assistance for Eastern Europe and the Baltic States and Assistance for the Independent States of the Former Soviet Union.

t. Development Assistance includes sub-account programmatic activities to support USAID national, regional, and central international climate change activities.

u. International Disaster Assistance includes sub-account programmatic activities to support USAID national, regional, and central international climate change activities.

v. Economic Support Fund includes sub-account programmatic activities to support USAID national, regional, and central international climate change activities.

w. Complementary Agencies (and their programmatic activities) include U.S. Department of Energy (including the China and India Clean Energy Research Centers); U.S. Environmental Protection Agency (including Methane for Markets, International Capacity Building, and contributions to the Multilateral Fund to support the Montreal Protocol on Substances that Deplete the Ozone Layer); U.S. Department of Commerce (including the National Oceanic and Atmospheric Administration's International Research Institute for Climate and Society and the

International Trade Administration's activities under the Asia Pacific Partnership); National Science Foundation (including the Regional Institutes for Global Change and the Basic Research to Enable Agricultural Development); U.S. Department of Agriculture (including international programmatic activities in the Forestry Service); National Aeronautics and Space Administration (including the SERVIR initiative); Millennium Challenge Corporation; and U.S. Trade and Development Agency; but do not include Overseas Private Investment Corporation or Export-Import Bank. GCCI funding was not reported for complementary agencies prior to FY2009.

End Notes

[1] See White House Fact Sheet: U.S. Global Development Policy at http://www.whitehouse.gov/the-press-office/2010/ 09/22/fact-sheet-us-global-development-policy.

[2] The GCCI was one of three main initiatives to the 2010 Global Development Policy that also included the Global Health Initiative and Global Food Security.

[3] For more information on this program, see CRS Report R41302, *International Climate Change Financing: The Climate Investment Funds (CIF)*, by Richard K. Lattanzio.

[4] For more information on this program, see CRS Report R41165, *International Environmental Financing: The Global Environment Facility (GEF)*, by Richard K. Lattanzio.

[5] "Incremental" funding refers to costs associated with transforming a project with national benefits into one with global environmental benefits. Global Environment Facility (GEF) grants aim to cover the difference or "increment" between a less costly, more polluting option and a costlier, more environmentally sound option. In this way, GEF funding is structured to "supplement" base project funding and provide for the environmental components in national development agendas.

[6] For more information on this program, see CRS Report R41302, *International Climate Change Financing: The Climate Investment Funds (CIF)*, by Richard K. Lattanzio.

[7] For more information on this program, see CRS Report R41302, *International Climate Change Financing: The Climate Investment Funds (CIF)*, by Richard K. Lattanzio.

[8] For more information on this program, see the United Nations Framework Convention on Climate Change website at http://unfccc.int/cooperation 3660.php.

[9] For more information on this program, see the United Nations Framework Convention on Climate Change website at http://unfccc.int/cooperation3657 .php.

[10] For more information on this program, see CRS Report R41302, *International Climate Change Financing: The Climate Investment Funds (CIF)*, by Richard K. Lattanzio.

[11] Estimate based on Division F, the Department of State, Foreign Operations, and Related Programs Appropriations Act, 2010 (P.L. 111-117) in which funding was $16.1 billion for the Department of State, International Broadcasting, and related agencies and $32.8 billion for Foreign Operations, for a total of $48.9 billion. FY2010 GCCI programming accounted for just over $1 billion. The GCCI FY2011 and FY2012 requests are comparable. Considering that the Department of State, Foreign Operations, and Related Programs budget generally accounts for approximately 3% of U.S. discretionary spending, the GCCI accounts for less than 0.1% of all U.S. discretionary spending. For more detail on recent Department of State, Foreign Operations, and Related Programs budgets, see CRS Report R41680, *Fact Sheet: The FY2012 State and Foreign Operations Budget Request*, by Susan B. Epstein, Marian Leonardo Lawson, and Tamara J. Resler; CRS Report R41228, *State, Foreign Operations, and Related Programs: FY2011 Budget and Appropriations*, by

Marian Leonardo Lawson, Susan B. Epstein, and Tamara J. Resler; and CRS Report R40693, *State, Foreign Operations, and Related Programs: FY2010 Budget and Appropriations*, by Susan B. Epstein, Kennon H. Nakamura, and Marian Leonardo Lawson.

[12] It should be noted that many GCCI activities are funded at agency sub-account levels, and thus, many allocations for these programmatic activities are left to the discretion of the agencies, under congressional consultation, and could change as offices, bureaus, and/or missions update priorities in their operational budgets.

[13] These figures do not reflect the 0.2% rescission across all non-defense accounts for FY2011 funds, in accordance with Section 1119(a) of P.L. 112-10.

[14] For more information on the negotiations, see CRS Report R40001, *A U.S.-Centric Chronology of the International Climate Change Negotiations*, by Jane A. Leggett.

[15] Jonathan Pershing, Deputy Special Envoy for Climate Change and Head of the U.S. Delegation, at a January 5, 2011, talk at the Center for Strategic and International Studies, said that the Copenhagen and Cancun conferences represent a paradigm shift from a top-down method to a bottom-up structure in which countries make emissions reduction pledges and move forward and take actions suitable to their national circumstances.

[16] William Easterly and Tobias Pfutze, "Where Does the Money Go? Best and Worst Practices in Foreign Aid," *Journal of Economic Perspectives*, vol. 22, no. 2 (Spring 2008). For more on foreign aid reform, also see CRS Report R40102, *Foreign Aid Reform: Studies and Recommendations*, by Susan B. Epstein and Matthew C. Weed; and CRS Report R40756, *Foreign Aid Reform: Agency Coordination*, by Marian Leonardo Lawson and Susan B. Epstein.

[17] Both the World Bank and U.S. Geological Survey estimate that every dollar spent on disaster preparedness saves $7 in disaster response. The World Bank, *Natural Disasters: Counting the Cost,* March 2, 2004, at http://go.worldbank.org/NQ6J5P2D10 (accessed May 20, 2011).

[18] National Intelligence Council, *National Intelligence Assessment on the National Security Implications of Global Climate Change to 2030, Statement for the Record by Dr. Thomas Fingar, Deputy Director of National Intelligence for Analysis, National Intelligence Council, before the U.S. Congress, House Permanent Select Committee on Intelligence & House Select Committee on Energy Independence and Global Warming,* June 25, 2008, at http://www.dni.gov/nic/ special_climate2030.html (accessed March 31, 2011).

[19] For more information on the political issues involved in the international climate change negotiations, see CRS Report R40001, *A U.S.-Centric Chronology of the International Climate Change Negotiations*, by Jane A. Leggett.

In: International Financing Efforts to Address... ISBN: 978-1-62100-702-9
Editors: Randy N. Matthews

Chapter 3

INTERNATIONAL CLIMATE CHANGE FINANCING: THE CLIMATE INVESTMENT FUNDS (CIF)

Richard K. Lattanzio

SUMMARY

The United States contributes funding to various international financial institutions to assist developing countries address global climate change and other environmental concerns. Congress is responsible for several activities in this regard, including (1) authorizing periodic appropriations for U.S. financial contributions to the institutions, and (2) overseeing U.S. involvement in the programs. Issues of congressional interest include the overall development assistance strategy of the United States, U.S. leadership in global environmental and economic affairs, and U.S. commercial interests in trade and investment. This report provides an overview of two of the larger and more recently instituted international financial institutions for the environment—the Climate Investment Funds (CIF)—and analyzes their structure, funding, and objectives in light of the many challenges within the contemporary landscape of global environmental finance.

The CIF are investment programs administered by the World Bank Group that aim to help finance developing countries' transitions toward low-carbon and climate-resilient development. Formally approved by the World Bank's Board of Directors on July 1, 2008, the CIF are composed of two trust funds—

the Clean Technology Fund (CTF) and the Strategic Climate Fund (SCF)—each with a specific scope, objective, and governance structure. The CTF provides financing for demonstrating, deploying, and diffusing low-carbon technologies that have the potential for long-term avoidance of greenhouse gas emissions. The SCF—a suite of three separate funds, including the Pilot Program for Climate Resilience (PPCR), the Forest Investment Program (FIP), and the Scaling Up Renewable Energy Program in Low Income Countries (SREP)—supports the least developed countries in their efforts to achieve low-carbon, climate-resilient development. Overall, donor countries have pledged $6.9 billion to the funds since September 2008 in support of programs in 45 developing countries. The U.S. pledge totals $2 billion. In FY2010, Congress approved $375 million for the CIF (P.L. 111-117, Consolidated Appropriations Act, 2010); in FY2011, Congress approved $235 million for the CIF (P.L. 112-10, Continuing Appropriations). For FY2012, the Administration has requested an additional $590 million for the program.

The CIF are just one set of financial mechanisms in a larger network of international programs designed to address the global environment. Accordingly, their effectiveness depends on how the trust funds address programmatic issues, build upon national investment plans, react to recent developments in the financial landscape, and respond to emerging opportunities. Proponents of the CIF point to several factors in support of the funds, including an innovative programmatic design, a country-led investment process, and a balanced governance structure with enhanced stakeholder engagement. Proponents of the multilateral development banks' (MDBs') role in environmental assistance emphasize several advantages to financing climate programs through the World Bank Group, including its commitment to private sector development, its capacity to leverage large co-financing arrangements, and its possession of fiduciary standards and institutional expertise. However, critics highlight several factors of concern with the CIF and their Trustee, including a lack of transparency, coordination, and "polluter pay" responsibilities; a potential for new conditionalities, additionalities, and increased debt burdens on developing countries; and a prior economic development policy at the World Bank that is considered a conflict of interest for environmental protection.

INTRODUCTION

Many governments acknowledge that environmental degradation and climate change pose international and trans-boundary risks to human populations, economies, and ecosystems that could result in a worsening of poverty, social tensions, and political stability. To confront these global challenges, countries have negotiated various international agreements to protect the environment, reduce pollution, conserve natural resources, and promote sustainable growth. While some observers call upon developed countries to take the lead in addressing these issues, efforts are unlikely to be sufficient without similar measures being implemented in developing countries. Developing countries, however, focused on poverty reduction and economic growth, do not have the financial resources, technological know-how, and institutional capacity to deploy such measures. Therefore, increased international support in these areas has remained the principal method for governments to assist developing country action on global environmental problems.[1]

The United States and other industrialized countries have committed to financial assistance for environmental initiatives through several multilateral agreements (e.g., the Montreal Protocol (1987), the United Nations Framework Convention on Climate Change (1992), United Nations Convention to Combat Desertification (1994), and the Copenhagen Accord (2009)). International financial assistance takes many forms, from fiscal transfers to market transactions, and includes foreign direct investment (FDI), bilateral overseas development assistance (ODA), and contributions to multilateral development banks (MDB)[2] and other international financial institutions (IFI), as well as the offering of export credits, loan guarantees, insurance products, etc.

Table 1 outlines recent U.S. financial support for global environmental initiatives.[3] Congress is responsible for several activities in this regard, including (1) authorizing periodic appropriations for U.S. financial contrib.-utions to the institutions, and (2) overseeing U.S. involvement in the programs. Issues of congressional interest include the overall development assistance strategy of the United States, U.S. leadership in global environ-mental and economic affairs, and U.S. commercial interests in trade and investment.[4] As Congress considers potential authorizations and/or appropriations for initiatives administered through the Department of State, the Department of the Treasury, and other agencies with international programs, it may have questions concerning the direction, efficiency, and effectiveness of current bilateral and multilateral programs. This report provides an overview of two of the larger and more recently instituted multilateral mechanisms—the Climate Investment Funds (CIF)—and analyzes their structure, funding, and objectives in light of the many challenges within the contemporary landscape of global environmental finance.

Table 1. Recent U.S. Budget Authority for Multilateral Environmental Programs In Nominal US$ Millions

Agency/Program	2010 Enacted	2011 Request	2011 Enacted	2012 Request
Department of State				
Least Developed Country Fund	30	30	TBD[a]	30
Special Climate Change Fund	20	20	TBD	20
World Bank Forest Carbon Partnership	10	15	TBD	21
Department of the Treasury				
Tropical Forests Conservation Act	20	20	TBD	15
Global Environment Facility	87	175	90	144
Clean Technology Fund	300	400	185	400
Strategic Climate Fund: Pilot Program for Climate Resilience	55	90	/ 50[b]	40
Strategic Climate Fund: Forest Investment Program	20	95	/ 50	130
Strategic Climate Fund: Scaling-Up Renewable Energy	0	50	/ 50	20

Source: Office of Management and Budget, *Federal Climate Change Expenditures Report to Congress*, 2010; Office of Management and Budget, *The Budget of the United States Government*, 2011 and 2012 (State, Foreign Operations and Related Programs Budget), and H.R. 1473, The Department of Defense and Full-Year Continuing Appropriations Act, 2011 (P.L. 112-10).

a. TBD, "to be determined": Appropriated funds for these specific programs/activities are drawn from larger line item categories in agency budget authorities, occasionally with "shall"-language implementing spending ceilings. Based on provisions in H.R. 1473 (P.L. 112-10), allocations in FY2011 for these programs are left at the discretion of the agency and have yet to be determined/reported.

b. H.R. 1473 (P.L. 112-10) FY2011 budget authority for the Strategic Climate Fund is $50 million for all programming. Allocation of funds among the three sub-programs is at the discretion of Treasury and has yet to be determined/reported.

THE CLIMATE INVESTMENT FUNDS

Background

Projected climate change is considered a potential threat to economic development, with anticipated effects on the environment, human health, food security, and economic activity. Further, climate change disproportionately affects the urban and rural poor of developing countries, thus making it a central concern to those interested in poverty reduction, and sustainable development.[5] Under this context, and at the request of the G8/G20, the multilateral development banks (MDBs) have recently sought to expand their support to low-carbon and climate-resilient investments in several ways, including (1) creating new and additional environmental funding resources, (2) repackaging their "core" financial products with specialized climate provisions, and (3) leveraging their suite of financial instruments for greater private sector environmental investment.[6]

In keeping with these aims, in February 2008, Japan, the United Kingdom, and the United States announced their intention to create a set of funds at the World Bank to help developing countries "bridge the gap between dirty and clean energy" and "boost the World Bank's ability to help developing countries tackle climate change."[7] The World Bank held the first design meeting for the proposed Climate Investment Funds (CIF) in March 2008 in Paris, France. Two subsequent meetings were held in Washington, DC, and Potsdam, Germany, and on May 23, 2008, representatives from 40 developing and industrialized countries reached agreement on the funds' design and duration (the CIF are programmed to sunset upon the commencement of a new climate fund in the UNFCCC). Formally approved by the World Bank's Board of Directors on July 1, 2008, the CIF have become an attempt to bridge the gap in climate financing between present obligations and a post-2012 global climate change agreement.[8]

The CIF are composed of two separate trust funds—the Clean Technology Fund (CTF) and the Strategic Climate Fund (SCF)—each with a specific scope, objective, and governance structure. Overall, donor countries have pledged $6.9 billion to the funds since September 2008 in support of programs in 45 developing countries.[9] The U.S. pledge totals $2 billion. In FY2010, Congress approved $375 million for the CIF (P.L. 111-117, Consolidated Appropriations Act, 2010); in FY2011, Congress approved $235 million for the CIF (P.L. 112-10, Continuing Appropriations). For FY2012, the Administration has requested an additional $590 million for the program.

All U.S. funding is subject to annual congressional approval. Authorizing legislation is managed by the House Financial Services Committee and Senate Foreign Relations Committee. The House and Senate Appropriations Subcommittees on State, Foreign Operations, and Related Programs have jurisdiction over appropriations.

The Clean Technology Fund (CTF)

Overview

Faced with energy and environmental challenges, among others, many developing countries see value in clean technology to meet their energy security, poverty alleviation, and sustainable development goals while also reducing their growth in emissions. However, the costs to developing countries of switching to cleaner technologies without financial assistance may be prohibitive. The CTF seeks to provide financing—principally to larger emerging economies and to regional groups—for demonstrating, deploying, and diffusing low-carbon technologies with the potential for long-term avoidance of greenhouse gas emissions. The fund promotes renewable energy and energy efficient technologies in the power sector as well as energy efficiency strategies in the transportation, building, industry, and agricultural sectors. Currently, the CTF is designed to support 15-20 country and regional investment plans and/or co-financed projects. As of December 2010, the CTF has endorsed 14 programs, including plans from Mexico, Egypt, Turkey, South Africa, Ukraine, Morocco, Colombia, Kazakhstan, Indonesia, Philippines, Thailand, Vietnam, Nigeria, and a five-country region in the Middle East and North Africa. Projects include support for wind energy, urban public transportation systems, solar water heaters, smart-grid development, and concentrating solar thermal power programs, among others (see **Table 3** for more detailed descriptions of the national investment plans).[10]

Governance

The CTF is implemented by the World Bank Group and governed by representatives from the donor and recipient countries. The role of governance for the CTF is to approve investment plans, programming, and the allocation of financial resources; and to provide guidance, performance evaluation, and reporting. It is further tasked with ensuring that the strategic orientation of the CTF is guided by the principles of the United Nations Framework Convention on Climate Change (UNFCCC). The organizational structure of the CTF is equally balanced between donor and developing countries. All decisions are made by consensus. Other international organizations, the private sector, and civil society representatives are included as observers. All observer roles are "active," allowing them to take the floor to make interventions, propose agenda items, and recommend experts. Observers do not vote during consensus decisions. The governance structure includes:

- The CTF Trust Fund Committee, which oversees and decides on the operations and activities of the CTF and includes (1) eight representatives from contributor countries (currently Australia, France, Germany, Japan, Spain, Sweden, United Kingdom, United States); (2) eight representatives from eligible recipient countries (currently Brazil, China, Egypt, India, Morocco, Nigeria, South Africa, Turkey); (3) a representative from the project recipient country (during deliberations on the investment plan, program, or project); (4) a representative of the World Bank; and (5) a representative for the other MDBs.
- The MDBs Committee, which facilitates collaboration, coordination, and the exchange of information, knowledge, and experience among MDBs partners.
- The Partnership Forum, which supports civil society engagement and includes representatives of donor and eligible recipient countries, MDBs, UN and UN agencies, Global Environment Facility (GEF), UNFCCC, Adaptation Fund, bilateral development agencies, NGOs, indigenous peoples, private sector entities, and technical experts.
- The Administrative Unit, which supports the work of the CIF, is housed in the World Bank Group's Washington, DC, offices.
- A Trustee (the World Bank), which holds in trust, as the legal owner and administrator, the funds, assets, and receipts that constitute the Trust Fund, pursuant to the terms entered into with the contributors.

Funding

In September 2008, 13 donor countries pledged over US$6.9 billion to finance the two CIF trust funds. The total amount pledged by the eight contributing countries to the CTF was US$4.405 billion as of September 30, 2010 (see **Table 2** for pledges and **Table 1** for U.S. Budget Authority). The funds are to be disbursed as grants, concessional loans, loan guarantees, and other risk management instruments.

Table 2. Total Pledges to the Clean Technology Fund (In millions US$, as pledged since September 2008, with exchange rates as of September 30, 2010)

Donor	Contribution Type[a]	Amount Pledged
Australia	Grant	$97
France	Loan	$277
Germany	Loan/Grant	$615
Japan	Grant	$1,112
Spain	Capital	$109
Sweden	Grant	$90
United Kingdom	Capital	$613
United States[b]	Grant	$1,492
Total		$4,405

Source: The CIF website at http://www.climateinvestmentfunds.org/

a. Donor contribution types include grants, loans, and equity, and describe in broad terms the general. requirements stipulated by the donors on their contributed funds. The U.S. government has historically contributed grant financing for reasons that include ease, ODA accounting practices, and flexible capital reflow provisions.

b. The total U.S. pledge to the CIF remains at $2 billion. Contributions across funds are extrapolated from current allocations.

Program Areas

The CTF is based on country and regional investment plans that aim to support climate-friendly technologies. Investment plans are undertaken jointly by the recipients, the MDBs, other development partners, private industry, and civil society to build upon existing national strategies and demonstrate how the CTF can be complementary to the country's overall developmental activities. The CTF supports investment plans that are cost-effective and implement-tation-ready, can be scaled up quickly to impact development, and have the potential for significant greenhouse gas emission reductions. To receive CTF funding, a country must be eligible for official development assistance (ODA) and have an active MDB program.

The majority of CTF funding supports programs that help shape demand side markets for technology diffusion. The fund's criteria for lending allow for all renewable and energy efficiency initiatives, as well as large-scale hydro-lectric power plants, natural gas plants, some forms of biofuels, power plant refits, and ultra-supercritical coal plants.[11] Funds are commonly targeted to support a variety of investment activities, including (1) direct purchase of technological goods and services; (2) direct investment into government infrastructure for transport or transmission modernization; (3) seed funds for financial intermediaries to incentivize clean technology lending; and (4) investment support and risk mitigation strategies for private sector entry into the market. In short, the CTF attempts to address the additional costs contained in lower-carbon energy investment such that it becomes a viable option to conventional fossil-fuel power generation. **Table 3** outlines the endorsed Investment plans as of June 1, 2010.

Table 3. Clean Technology Fund Investment Plans (In millions US$)

Date[a]	Country	Direct CTF Funding/ Co-Financing[b]	Investment Plan
January 2009	Egypt	$300 / $1,600	Wind power; Urban transport (natural gas buses and a subway); Transmission upgrades.
January 2009	Mexico	$500 / $6,000	Energy efficiency (appliance & lighting); Urban transport (rapid bus); Wind power.
January 2009	Turkey	$250 / $1,900	Renewable energy and energy efficiency; Smartgrid technology.
October 2009	Morocco	$150 / $1,500-$1,800	Energy sector privatization; Energy conservation; Urban transport.
October 2009	South Africa	$500 / $1,900	Concentrated solar power; Wind power; Solar water heaters; Energy efficiency.
October 2009	M.E.N.A.[c]	$750 / $4,900	Concentrated solar power; Transmission and distribution infrastructure.
December 2009	Thailand	$300 / $4,000	Renewable energy and energy efficiency; Urban transport (bus system).
December 2009	Philippines	$250 / $2,500	Solar power; Transmission infrastructure; Demand side management; Sustainable transport strategy.
December 2009	Vietnam	$250 / $3,200	Renewable energy and industrial energy efficiency; Urban transport (rail system); Initial capitalization of funds; Transmission infrastructure.
March 2010	Colombia	$150 / $3,000	Sustainable transport program; Public/private sector energy efficiency program.

Table 3. (Continued)

Date[a]	Country	Direct CTF Funding/ Co-Financing[b]	Investment Plan
March 2010	Indonesia	$400 / $2,700	Large-scale geothermal power; Biomass and other renewable energy.
March 2010	Kazakhstan	$200 / $535	Hydro and wind power; Public sector transport fuel switch; Dist-rict heating; Energy efficiency.
March 2010	Ukraine	$350 / $2,300	Wind, hydro, biomass; Residential and government energy efficiency; District heating; Smartgrid technology.
November 2010	Nigeria	$250 / $1,300	Transport sector structure; Clean and renewable energy development; Energy efficiency; Financial sector reform.

Source: CTF committee meeting documents and national Investment plans, available at the CTF website.

a. Date of official CTF endorsement of the investment plan.

b. Endorsed funding by the CTF / leveraged co-financing from additional sources; in millions US$. It is estimated that 30% of the co-financing comes from the private sector; the remainder comes from other multilateral financial institutions, the recipient governments, state-owned enterprises, and carbon finance.

c. The Middle East North Africa region, including Algeria, Egypt, Jordan, Morocco, and Tunisia.

The Strategic Climate Fund (SCF)

Overview

Some governments and civil society organizations are concerned that climate change may exacerbate poverty situations and reverse economic gains in the developing world through the possibility of temperature increases, rising sea levels, droughts, changes in rainfall patterns, heightened disease patterns, and the lack of drinkable water. They believe that resources may be necessary to help low-income countries manage a response. Responses to climate change are likely to entail both mitigation efforts (i.e., slowing, then reducing greenhouse gas emissions) and adaptation efforts (i.e., managing the effects of short- and long-term climate outcomes). The SCF aims to help developing countries prepare for climate change by promoting low-carbon, climate-resilient development. Three targeted programs provide grants and concessional loans to pilot new approaches aimed at specific challenges:[12]

- The Pilot Program for Climate Resilience (PPCR) supports ways to integrate climate risk and resilience into the development strategies of low-income countries. Funds can be used to provide technical assistance to help with capacity building, policy reform, and sector investment.
- The Forest Investment Program (FIP) provides financing to countries to help them prepare for and participate in programs that aim to reduce deforestation. Funds can be used for managing forests and for educating indigenous and local communities about forest policies.
- The Scaling Up Renewable Energy Program in Low Income Countries (SREP) helps low-income countries adopt renewable energy solutions to aid in the development of their power generation sector. Funds can be used to provide policy support, technical assistance, financial management, and sector investment.

Governance

The SCF is implemented by the World Bank Group and governed by representatives from the donor and recipient countries. The governance and decision-making structure is similar to the CTF, but specifically includes:

- The SCF Trust Fund Committee, which oversees and decides on the operations and activities of SCF and includes (1) eight representatives from contributor countries (currently Australia/UK (rotating), Canada, Denmark/Switzerland (rotating), Germany, Japan, Netherlands, Norway, United States); (2) eight representatives from eligible recipient countries (currently Bolivia, Guyana, Indonesia, Kyrgyz Republic, Maldives, Senegal, Tunisia, Yemen); (3) a representative of the World Bank; and (4) a representative for the other MDBs.
- An SCF sub-committee for each of the targeted programs, which includes up to six representatives from contributor countries to the SCF Program, a matching number of representatives from eligible recipient countries, and such other representatives designated by the SCF Trust Fund.

Table 4. Total Pledges to the Strategic Climate Fund (In millions US$, as pledged since September 2008, with exchange rates as of September 30, 2010)

Donor	Contribution Type[a]	Amount Pledged
Australia	Grant	$48
Canada	Grant	$84
Denmark	Grant	$35
Germany	Loan/Grant	$68
Japan	Grant	$222
Netherlands	Grant	$76
Norway	Grant	$179
Switzerland	Grant	$20
United Kingdom	Capital	$1,273
United States[b]	Grant	$508
Total		$2,514

Source: The CIF website at http://www.climateinvestmentfunds.org/

a. Donor contribution types include grants, loans, and equity, and describe in broad terms the general. requirements stipulated by the donors on their contributed funds. The U.S. government has historically contributed grant financing for reasons that include ease, ODA accounting practices, and flexible capital reflow provisions.

b. The total U.S. pledge to the CIF remains at $2 billion. Contributions across funds are extrapolated from current allocations.

- The MDBs Committee, which facilitates collaboration, coordination, and the exchange of information, knowledge, and experience among the MDBs partners.
- The Partnership Forum, which supports civil society engagement and includes representatives of donor and eligible recipient countries, MDBs, UN and UN agencies, GEF, UNFCCC, Adaptation Fund, bilateral development agencies, NGOs, indigenous peoples, private sector entities, and technical experts.
- The Administrative Unit, which supports the work of the CIF, is housed in the World Bank Group's Washington, DC, offices.
- A Trustee (the World Bank), which holds in trust, as the legal owner and administrator, the funds, assets, and receipts that constitute the Trust Fund, pursuant to the terms entered into with the contributors.

Funding

In September 2008, 13 donor countries pledged over US$6.9 billion to finance the two CIF trust funds. The total amount pledged by 10 countries to the SCF was US$2.514 billion as of September 30, 2010 (see Table 4 for pledges and Table 1 for U.S. Budget Authority). The funds are to be disbursed as grants, concessional loans, loan guarantees, and other risk management instruments.

Program Areas

The programming of the SCF is less advanced than that of the CTF. Each of the three funds remains in early implementation stages with its trust fund committee, having launched no earlier than January 2009. As of the November 2010 meeting of the Joint CTF and SCF Trust Fund Committees, the status of each fund was reported as follows:[13]

- The Pilot Program for Climate Resilience: The PPCR became operational in January 2009; $972 million of the SCF pledge has been targeted to the PPCR. Two regional groupings—the Pacific and the Caribbean—and nine countries, including Bangladesh, Bolivia, Cambodia, Mozambique, Nepal, Niger, Tajikistan, Yemen, and Zambia, were invited to participate. The program provides funding to the countries in two phases: (1) a technical assistance phase, which includes looking at how countries' development plans can be made more climate-resilient and deciding upon the types of investments countries could make, and (2) an implementation phase, which includes the dispersal of grants of up to $1.5 million with the option of additional loans to implement programs. Zambia was the first country to begin preparations for phase one. Zambia's process was expected to raise discussion in the PPCR sub-committee regarding deliverables to guide other countries. Cambodia, Tajikistan, Yemen, Mozambique, and Bolivia followed in November 2009 in presenting proposals for development. All PPCR pilot countries convened in November 2010 to continue building a community of practice, exchange experiences, and document good practices from the design process of PPCR.
- The Forest Investment Program: With a February 2010 pledge of $67 million from Japan, the FIP was operational with a total of $558 million targeted to the program. The next steps were to select five pilot countries and three back-up countries; to finalize and approve

operational guidelines, financing, and investment criteria; and to design a grant mechanism to support the participation of indigenous peoples and local communities. In March 2010, the FIP subcommittee approved pilot programs in Indonesia, Ghana, Laos, Peru, and Burkina Faso; with Brazil, Democratic Republic of Congo, and Mexico approved subsequently in June 2010. All eight pilot country governments have confirmed in writing their interest in being supported under the FIP. Work has been initiated in each. The Multilateral Development Banks (MDBs) have reached out informally to the government representatives in the pilot countries and are in the process of planning jointly with the governments in partnership with country stakeholder groups and other interested partner institutions. The first meeting of pilot countries for the FIP took place in November 2010. The objective of the meeting was to build a community of practice among the FIP pilot countries to exchange experiences, to document good practices from the design process of Investment Strategy, and to reach a shared understanding provided by the FIP.

- The Scaling Up Renewable Energy Program in Low Income Countries: The SREP become operational in December 2009, surpassing its target of $250 million with a pledge of $50 million from the United States at the Copenhagen Conference of Parties to the UNFCCC. With a February 2010 pledge of $40 million from Japan, and a June 2010 pledge of $11 million from Denmark, the fund now has a total of $307 million. The SREP sub-committee held meetings in early February 2010 to set criteria for selecting an advisory group as well as pilot countries. In June, Ethiopia, Honduras, Kenya, Maldives, Mali, and Nepal were selected. All six pilot country governments have confirmed in writing their interest in being supported under the SREP. The committee aimed to approve programming procedures, financing, and a measurement framework over the course of 2010. The first scoping and joint missions are expected to be organized for early 2011 once the programming and financing modalities have been approved by the sub-committee. In many cases, the MDBs have already reached out informally to the government representatives in the pilot countries to begin preliminary discussions on SREP activities.

CURRENT ISSUES

Each year, billions of dollars in environmental aid flow from developed country governments—including the United States—to developing ones.[14] While the efficiency and the effectiveness of these programs are of concern to donor country governments, a full analysis of the purposes, intents, results, and consequences behind these financial flows has yet to be conducted.[15] International relations, comparative politics, and developmental economics can often collide with global environmental agendas. Critics contend that the existing system has had limited impact in addressing major environmental concerns—specifically climate change and tropical deforestation—and has been unsuccessful in delivering global transformational change. A desire to achieve more immediate impacts has led to a restructuring of the MDBs' role in environmental finance and the introduction of many new bilateral and multilateral funding initiatives. The CIF grew out of these concerns.

The effectiveness of the CIF depends on how the trust funds address their programmatic issues, build upon their national investment plans, react to recent developments in the financial landscape, and respond to emerging opportunities. The following section investigates some of the current challenges facing the CIF and summarizes some of the responses initiated by the funds.

Innovations by the CIF

Since their inception, the CIF have attempted to provide innovative approaches to global environmental issues and have introduced several processes to address the limitations of previous environmental finance.[16] These innovations include, but are not limited to, the following:

- *Programmatic Design*: While the CIF still aim to scale up existing practices and fund activities at the project level, they also were created to serve as laboratories for new financing schemes and vehicles for developing sustainable strategies. Funding strives to target the potential for large-scale transformation and to attain global environmental benefits. Stakeholders seek to share knowledge gained and inspire the use of best practice. As such, multinational or regional investment plans that support global development goals, energy security, industrial growth, diversification, and regional integration (e.g., the M.E.N.A. plan) best exemplify the CIF's programmatic approach.

- *Country-led Process*: Beyond a simple project-by-project approach, the purpose of the CIF is to bolster the efforts of countries' official adaptation plans and their actions toward low-carbon, climate-resilient development. The country-led approach aims to integrate funding into the country-owned development strategies consistent with the Paris Declaration.[17]
- *Innovative Governance and Stakeholder Engagement*: In an effort to attain transparency and accountability, the governing structure of the CIF is equally balanced between donor and developing countries. All decisions are taken by consensus, with no provision for voting. If a consensus is not possible, the proposal is postponed or withdrawn. Representatives from other international organizations, the private sector, and civil society are included as observers. All observer roles are "active," allowing them to take the floor to make interventions, propose agenda items, and recommend experts.

Issues in Support of the Multilateral Development Banks (MDBs) and Multilateral Assistance

The choice of financial mechanism and its administration is an important element to environmental finance. The differences among multilateral or bilateral assistance, grant or lending institutions, regional or global organizations, etc., all play a role in the structure of assistance. The decision to employ the MDBs as trustees for the CIF has both advantages and disadvantages. Historically, the MDBs have provided financial assistance to developing countries, typically in the form of loans and grants, for investment projects and organizational capacity.[18] Donor country support for the MDBs—including U.S. support—has assisted efforts to promote institutions, strengthen financial systems, undertake large infrastructure and social welfare projects, and develop property rights and rules of law. Through increased global integration, the aim of the MDBs has been to bolster economic growth, poverty alleviation, and resource allocation (including greater access to electricity) in developing countries while simultaneously building new markets for developed countries' exports and jobs. In 2008, at the urging of some donor countries,[19] a strategy to address climate change was added to the MDBs' development agenda. The "Strategic Framework on Development and

Climate Change"[20] analyzed the risks of climate change to economic development and served as a basis for integrating mitigation and adaptation planning into national development plans. Donor countries see several advantages to financing climate programs through the institutional structure of the MDBs. These advantages include, but are not limited to, the following:

- *Commitment to Private Sector Development*: Many donor countries—including the United States—believe that climate-friendly economic growth can be led by the private sector through such efforts as improving access to financial markets, building the capacity of entrepreneurs, and providing training to civilian society. One aim of the MDBs is to help foster private sector development by leveraging donor funds into highly effective co-financing arrangements. Historically, the U.S. Administration has supported these efforts. In a March 25, 2010, hearing before the House Appropriations Subcommittee on State, Foreign Operations, and Related Programs, the Treasury Department went on record as stating that the United States invests in the MDBs because "they help generate new engines of growth that benefit the U.S. economy and the global economy as a whole."
- *Economies of Scale, Coordination, and Co-Financing*: Proponents of the MDBs argue that multilateral assistance can solve problems of scale and efficiency by providing specialized expertise while lowering administration and coordination costs. Similarly, more competitive procurement rules, attractive cost-sharing opportunities, and the ability to leverage co-financing from other public and private organizations allow the MDBs to play a catalytic role in mobilizing financial aid.[21] At the March 25, 2010, hearing noted above, the Treasury Department stated that the MDBs "provide strong, effective and highly leveraged means to advance global prosperity.... For every dollar the United States contributes to paid-in capital for the World Bank, six dollars of additional capital is generated by other donors. And, for every dollar we invest in the World Bank, $26 worth of aid is delivered."[22]
- *Responsiveness to Donors*: The Treasury Department has similarly stated that the United States invests in the MDBs because they "promot[e] core American interests and values." This arrangement is due primarily to the structure and organization of the banks. MDBs governance is weighted on the basis of the cumulative financial

contributions and commitments by the donor countries, and thus, while a single trust fund, like the CIF, may be designed to balance equally the roles of developed and developing countries, the MDBs are designed to give greater weight to the major donors. The United States retains the most influence on World Bank matters, with a 16.4% voting share and the ability to veto major policy decisions. It is followed by Japan in second place, Germany in fourth, and France and the United Kingdom tied for fifth. The only developing or emerging country with as much voting interest is China, at third, with 4.4%.[23] With a governing structure that requires one representative from the World Bank, as trustee, and one representative from the group of remaining MDBs, as well as eight representatives from participating donor countries, the overall governance structure of the CIF has remained responsive to donor interests.

- *Possession of Fiduciary Standards*: Both current and past Administrations have argued that the World Bank has the proper internal safeguards to oversee large amounts of financing. As reported by the Department of Treasury, "the World Bank is an attractive trustee [for environmental funds] precisely because of its strong fiduciary standards and its extensive capacity to uphold them."[24]
- *Possession of Institutional Expertise, Information, and Credibility Provisions*: Proponents of the MDBs claim that multilateral agencies offer larger and better trained staffs with greater technical expertise. They state that large infrastructure investment, particularly in innovative technologies and methods, requires professionals who are experienced in identifying and facilitating access to technology, sharing risks associated with commercialization, and improving institutional capacity. Beyond institutional knowledge, multilaterals also collect, interpret, and disseminate costly information on a global scale and provide credibility controls for both recipient and donor governments.

Issues of Concern for Developing Countries and Civil Society

While advantages exist to financing climate programs through the institutional structure of the MDBs, concerns also persist. A variety of recipient countries, nongovernmental organizations, and civil society groups[25] have highlighted a number of issues, including, but not limited to, the following:

- *Donor Centrism*: While efforts have been made to include developing countries in the governance of the CIF, some commentators still point to the fact that the initiative was led by Japan, the United States, and the United Kingdom, and that it existed outside international climate change regulatory frameworks and without the initial participation of developing countries and other stakeholders. Moreover, they claim that the design of the CIF remains premised on an aid framework that is based on a donor-donee relationship that runs contrary to the international climate change principles of developed country obligations and "common but differentiated responsibilities."[26]
- *Lack of Transparency*: While efforts have been made at transparency and accountability, some sessions of the CIF committee meetings remain closed to observers. Similarly, technology investment plans are not publicly disclosed due to claims about sensitive sovereign information and national priorities. Observers from GEF and UNFCCC are also excluded from investment plan discussions, making it difficult to ensure complementarity.[27]
- *Lack of Coordination*: Observers point out that the CIF may create a parallel structure for financing climate change adaptation and mitigation efforts outside the ongoing multilateral framework for climate change negotiations. They are concerned that without harmonization between the CIF and the other sources of environmental finance (e.g., funds managed by the UN, the Global Environment Facility, and bilateral sources), overlaps, redundancies, competing views, and lack of synergy may affect climate priorities, funding processes, and qualifying criteria.
- *Potential to Prejudice U.N. Climate Provisions*: Some commentators and several governments have expressed concerns that the establishment of the CIF as trust funds to the MDBs may prejudice the outcomes of the international negotiations on climate finance within the framework of the United Nations. Many developing countries have expressly stated that they do not consider funds contributed to the CIF as meeting U.N. Annex I obligations. Furthermore, the design of the CIF includes a "sunset clause" stating that the CIF "will take necessary steps to conclude its operations once

a new [UNFCCC] financial architecture is effective." The nature of these steps has yet to be determined.

- *Potential for New Conditionalities*: Under the CIF, individual loans or grants under country programs follow the investment lending policies and procedures of the MDBs, including their fiduciary standards and environmental and social safeguards. Some observers believe this burdens developing countries with the World Bank's traditional criteria for financing, including tight fiscal discipline and the implementation of economic and other structural and policy reforms. They fear that the CIF may create onerous obligations on developing countries to comply with emission reduction targets and other rules from which they have been previously exempt.
- *Lack of Polluter Responsibility*: Some commentators claim that the provision of loans—in addition to grants—as a financial instrument to eligible developing countries contradicts the internationally agreed principle of "polluter pays" as stated in the Rio Declaration. Some argue that the repayment of a loan, notwithstanding the degree of concessionality, burdens a developing country with self-paying for a problem (climate change) that was caused by others (i.e., developed countries).
- *Increased Debt Burden*: While the loans under the CIF are provided on a concessional[28] basis, they are co-invested in other projects and programs and will have to be repaid. Some commentators believe that the CIF's loans may add to their debt burden in the long run and affect their ability to generate resources for growth.
- *Potential for Additionality*: The UNFCCC provides that developed country signatories to the Convention "provide new and additional financial resources to meet the agreed full costs incurred by developing country Parties" in their efforts at mitigation and adaptation.[29] Some observers fear that the design of the CIF establishes a parallel process for climate financing that does not result in new and additional resources. They are concerned that significant portions of the aid budgets of donors may be diverted into the CIF and counted as part of their annual ODA commitments.
- *Commercial Influence*: While advantages exist in prioritizing market-based solutions to dealing with the problems of climate, some groups express concern that the private sector may be unduly driven by commercial interests at the expense of social or environmental safeguards.[30] Concern also exists that a dependence on market

mechanisms as a source of climate financing may be inadequate and inconsistent for meeting the financial needs of developing countries charged with the responsibility of both implementing climate change commitments and mediating the social, economic, and environmental dislocations brought on by climate change.

- *Energy Policy at the Banks*: Many observers claim that the history of the World Bank's energy and infrastructure lending undermines its credibility as an institution committed to combating the impacts of climate change. Civil society groups have often highlighted the inconsistencies between the Bank's rhetoric on climate change and its operational policies and practices. They emphasize that while the Bank has increased financing for renewable energy and energy efficiency in recent years, its fossil fuel lending still accounts for 54% of the energy sector share for fiscal years 2006 to 2008 (compared to 10% for renewable energy, 15% for energy efficiency, and 21% for large hydropower). Furthermore, the trend has reportedly increased, as FY2008 has seen the World Bank and IFC scale up funding for fossil fuels by 102%, compared with only 11% for new renewable energy.[31] The controversy is compounded by the Bank's inability to reach a consensus on the definition of "clean energy technology," retaining provisions for ultra-supercritical coal-fired power generation in its environmental strategies.[32] Recent guidance from the U.S. Administration regarding the World Bank's engagement with coal-fired power generation in developing countries similarly leaves the definition open, stating that projects "could include more carbon efficient fossil fuel generation" in their portfolio.[33] While observers generally agree that funding from the CIF is unlikely to be used in coal-fired power generation projects, most agree that continued investment by the World Bank in fossil fuel energy and infrastructure may have several unintended effects, including, inter alia, (1) counteracting any gains made with the Bank's renewable portfolio, (2) directing resources toward large-scale power generation for industrial use rather than energy access and poverty reduction in poor urban and rural communities, and (3) drawing the Bank's professional and technical staff away from a concentration on energy efficiency and renewable energy activities to remain involved with fossil fuel generation.[34]

End Notes

[1] For a more detailed discussion on various sources and mechanisms of financial assistance for climate change activities, see CRS Report R41808, *International Climate Change Financing: Needs, Sources, and Delivery Methods*, by Richard K. Lattanzio and Jane A. Leggett.

[2] The group of multilateral development banks referred to in this report includes the International Bank for Reconstruction and Development (IBRD or World Bank), African Development Bank (AfDB), Asian Development Bank (ADB), European Bank for Reconstruction and Development (EBRD), Inter-American Development Bank Group (IDB), and the International Finance Corporation (IFC, the private sector wing of the IBRD).

[3] Some commentators believe the new and increased funding for environmental issues is the result of several factors, including (1) an increased political understanding by some policymakers of climate change, (2) the transformed role of multilateral development banks in global energy and environmental issues, (3) an expressed desire to achieve more immediate environmental and economic impacts through bilateral and private sector resources, and (4) a perceived lack of efficiency in current financial mechanisms. See Gareth Porter, Neil Bird, Nanki Kaur, and Leo Peskett, "New Finance for Climate Change and the Environment," WWF and the Heinrich Böll Foundation, 2008.

[4] For more substantive analysis of foreign aid and congressional roles, see CRS Report R40213, *Foreign Aid: An Introduction to U.S. Programs and Policy*, by Curt Tarnoff and Marian Leonardo Lawson; and CRS Report R41170, *Multilateral Development Banks: Overview and Issues for Congress*, by Rebecca M. Nelson.

[5] As summarized by the International Institute for Sustainable Development, at http://www.iisd.ca/download/pdf/sd/ ymbvol172num2e.pdf.

[6] See the World Bank website for additional information at http://siteresources.worldbank.org/NEWS/Resources/ Climate_Change_Results_Brief_4-12-10.pdf.

[7] Henry Paulson, Alistair Darling, and Fukushiro Nukaga, "Financial bridge from dirty to clean," *Financial Times*, February 7, 2008.

[8] For a full description of purpose and programs, see CIF's website at http://www.climateinvestmentfunds.org/cif/.

[9] Exchange rates as of September 30, 2010, the last recorded Trustee Report for the CIF, at http://www.climateinvestmentfunds.org/cif/sites/climateinvestmentfunds.org/files/ CTF%204%20Trustee%20Report%20nov2010.pdf.

[10] Description of CTF overview and governance from CIF, Annual Report 2009, on the CIF website at http://www.climateinvestmentfunds.org/cif/sites/climateinvestmentfunds.org/files/ cif_annual_report_final_021810.pdf; furthermore, for a more detailed background into the issues surrounding the implementation of the CTF, see CRS Report RS22989, *The World Bank's Clean Technology Fund (CTF)*, by Martin A. Weiss.

[11] No coal-fired power plants have been proposed or approved at this time. Hydroelectric power generation is currently included in the Ukraine proposal.

[12] Description of SCF overview and governance from CIF, Annual Report 2009, on the CIF website at http://www.climateinvestmentfunds.org/cif/sites/climateinvestmentfunds.org/files/cif_annual_report_final_021810.pdf.

[13] See SCP Committee document, "SCF/TFC.6/3, Progress Report on targeted programs under the SCF," at http://www.climateinvestmentfunds.org/cif/sites/climateinvestmentfunds.org/files/ SCF%203%20Progress%20Report%20on%20targeted%20programs%20nov2010.pdf.

[14] The Organisation for Economic Co-operation and Development (OECD) maintains information on Member countries' Official Development Assistance. Current data (accessed April 15, 2011) reflect that all OECD DAC Member countries contributed, on average, a total of $2,283 million per annum over the period 2005-2009 to multi-sectoral

environmental protection assistance (in 2010 US$), and that the United States contributed, on average, $285 million per annum over the same period to multi-sectoral environmental protection assistance (in 2010 US$). See OECD StatExtracts database at http://stats.oecd.org/Index. aspx?DataSetCode=ODA_DONOR#.

[15] This report does not aim to unpack the full range of discussions on environmental and developmental assistance. For a discussion on international development assistance in general, see CRS Report R40213, *Foreign Aid: An Introduction to U.S. Programs and Policy*, by Curt Tarnoff and Marian Leonardo Lawson. An overview and analysis of the history of environmental financing can be found in a number of source materials including recent book length studies by Inge Kaul and Pedro Conceição, *The New Public Finance: Responding to Global Challenges,* New York: Oxford University Press, 2006; and Robert L. Hicks, Bradley C. Parks, J. Timmons Roberts, and Michael J. Tierney, *Greening Aid?: Understanding the Environmental Impact of Development Assistance,* New York: Oxford University Press, 2008.

[16] For further discussion regarding the limitations of past mechanisms for global environmental finance, see the section on institutional challenges in CRS Report R41165, *Global Environment Facility (GEF): An Overview*, by Richard K. Lattanzio.

[17] The 2005 Paris Declaration, endorsed by over 100 countries, aims to increase harmonization, alignment, and management of aid for results with a set of actions and indicators that can be monitored. See http://www.oecd.org/ dataoecd/11/41/34428351.pdf.

[18] For a fuller discussion on the structure and the role of the MDB system, refer to CRS Report R41170, *Multilateral Development Banks: Overview and Issues for Congress*, by Rebecca M. Nelson.

[19] Including the United States. See the negotiations at the 2005 G8 Gleneagles Summit at http://www.g7. utoronto.ca/ summit/2005gleneagles/.

[20] See http://siteresources.worldbank.org/EXTCC/Resources/407863-1219339233881/DCCSFT echnicalReport.pdf.

[21] Sources of additional funds most often include other MDBs and multilateral financial institutions, the recipient governments, state-owned enterprises, and carbon finance, as well as the private sector.

[22] See testimony at http://appropriationsGeithner. 3.25.10.pdf.

[23] As reported by Reuters, at http://www.reuters.com/article/idUSTRE63O1RQ20100425, after the World Bank Development Committee spring meetings on April 25, 2010. See World Bank press release at http://web.worldbank.org/WBSITE/EXTERNAL/NEWS/ 0,,content MDK:22556192~pagePK:34370~piPK:34424~theSitePK:4607,00.html.

[24] As reported in Climatewire, "Eskom fallout spurs new opposition to World Bank's role in climate funding," May 24, 2010.

[25] There are many published critiques on the environmental agenda of the MDBs. Of specific relevance for CIF, see, for example, Celine Tan, "No Additionality, New Conditionality: A Critique of the World Bank's Proposed Climate Investment Funds," TWN, 2008, at http://www.twnside.org.sg/bangkok.briefings.htm; Smita Nakhooda, "Catalyzing Low-carbon Development?" WRI, 2009, at http://pdf.wri.org/working_papers/ development_ clean_technology_fund.pdf; and Heike Mainhardt-Gibbs, et al., "Fuelling Contradictions: The World Bank's Energy Lending and Climate Change," Bretton Woods Project, CRBM & URGEWALD, 2010, at http://www.brettonwoodsproject.org/art-566198.

[26] See UNFCCC, Art. 3.1, at http://unfccc.int/essential_background/convention items/1355.php.

[27] It should be noted that the United States, as well as the United Kingdom and France, have expressed a desire for some advance disclosure of investment plans and for observer participation in these sessions.

[28] "Concessional" or "soft" loans are loans extended on terms substantially more generous than market loans. The concessionality is achieved either through interest rates below those available on the market or by extended grace periods, or a combination of these.

[29] See UNFCCC, Article 4:3, at http://unfccc.int/essential_background/conventionbackground/items/1362.php.

[30] This concern has been levied against the Bank's brokering of carbon purchases through its Prototype Carbon Fund. See Bank Information Center, et al., "How the World Bank's Energy Framework Sells the Climate and Poor People Short," 2006, at http://www.bicusa.org/ en/Article.2954.aspx.

[31] See Bank Information Center report at http://www.bicusa.org/en/Article.11033.aspx.

[32] Ultra-supercritical coal-fired power generation is defined as "new pulverised coal combustion systems ... [that] operate at increasingly higher temperatures and pressures and therefore achieve higher efficiencies than conventional PCC units and significant CO2 reductions. Supercritical steam cycle technology has been used for decades and is becoming the system of choice for new commercial coal-fired plants in many countries." See World Coal Institute website at http://www.worldcoal.org/coal-the-environment/coal-use-the-environ ment/ improving-efficiencies/.

[33] See U.S. Treasury memorandum at http://www.ustreas.gov/offices/international-affairs/multilateral_banks/statements/COAL%20GUIDELINES%202009%2012%2014%20 FINAL%20%282%29.pdf.

[34] For discussion of further debate on this issue, see CRS Report RS22989, *The World Bank's Clean Technology Fund (CTF)*, by Martin A. Weiss; as well as the World Bank's issue brief on "Energy," available at http://web.worldbank.org/WBSITE/EXTERNAL/NEWS/0,,contentMDK:21513875~menuPK:34480~pagePK:64257043~piPK:437376~theSitePK:4 607,00.html; and, as one example, Heike Mainhardt-Gibbs, et al., "Fuelling Contradictions: The World Bank's Energy Lending and Climate Change," the Bretton Woods Project, CRBM & URGEWALD, 2010, at http://www. brettonwoodsproject.org/ art-566198.

In: International Financing Efforts to Address... ISBN: 978-1-62100-702-9
Editors: Randy N. Matthews

Chapter 4

INTERNATIONAL CLIMATE CHANGE FINANCING: THE GREEN CLIMATE FUND (GCF)

Richard K. Lattanzio

SUMMARY

Over the past several decades, the United States has delivered financial and technical assistance for climate change activities in the developing world through a variety of bilateral and multilateral programs. The United States and other industrialized countries committed to such assistance through the United Nations Framework Convention on Climate Change (UNFCCC, Treaty Number: 102-38, 1992), the Copenhagen Accord (2009), and the UNFCCC Cancun Agreements (2010), wherein the higher-income countries pledged jointly up to $30 billion of "fast start" climate financing for lower-income countries for the period 2010 – 2012, and a goal of mobilizing jointly $100 billion annually by 2020. The Cancun Agreements also proposed that the pledged funds are to be new, additional to previous flows, adequate, predictable, and sustained, and are to come from a wide variety of sources, both public and private, bilateral and multilateral, including alternative sources of finance.

One potential mechanism for mobilizing a share of the proposed international climate financing is the UNFCCC Green Climate Fund (GCF), currently under negotiation by Parties to the Convention. If established, the fund would be capitalized by contributions from donor countries and other sources and used to support climate change mitigation and adaptation projects, programs, policies, and other activities. The GCF would complement, or perhaps replace, many of the existing multilateral climate change funds (e.g., the Global Environment Facility, the Climate Investment Funds, the Adaptation Fund), and become the official financial mechanism of the Convention. A UNFCCC-appointed Transitional Committee has been tasked with designing the CGF, with the intent of bringing a finished proposal for a decision before the UNFCCC 17th Conference of Parties in Durban, South Africa, November 28 – December 9, 2011. Many issues remain to be clarified during negotiations, and some involve long-standing and contentious debate. They include: what role the CGF would play in providing sustained finance at scale, how it would fit into the existing development assistance and climate financing architecture, how it would be legally and institutionally governed, how it would be capitalized, and how it would allocate and deliver assistance efficiently and effectively to developing countries.

The U.S. Congress—through its role in authorizations, appropriations, and oversight—would have significant input on U.S. participation in the fund. As negotiations proceed, Congress may raise concerns regarding the cost, purpose, direction, efficiency, and effectiveness of the UNFCCC and existing international financial institutions. These concerns may be weighed against the negotiated design characteristics of the new fund in an effort to assess its potential performance. Congress may then be required to determine and give guidance to the allocation of funds between bilateral and multilateral climate change assistance as well as among the variety of multilateral mechanisms. Potential authorizations and appropriations for the GCF would rest with several committees, including the U.S. House of Representatives Committees on Foreign Affairs (various subcommittees); Financial Services (Subcommittee on International Monetary Policy and Trade); and Appropriations (Subcommittee on State, Foreign Operations, and Related Programs); and the U.S. Senate Committees on Foreign Relations (Sub-committee on International Development and Foreign Assistance, Economic Affairs, and International Environmental Protection); and Appropriations (Subcommittee on State, Foreign Operations, and Related Programs).

INTRODUCTION[1]

Many voices, domestic and international, have called upon the United States and other industrialized countries to increase foreign assistance to lower- and middle-income countries to address climate change.[2] Proponents maintain that such assistance could help promote climate-friendly and high-growth economic development in these countries, while simultaneously protecting the more vulnerable nations from the effects of a changing climate. For their part, most, if not all, lower-income countries have stated that their success at combating climate change would depend critically on receipt of international financial support. They argue that mitigating climate change pollutants, adapting to the effects of climate change, and building climate resilience into their development agendas would incur costs above and beyond their normal economic growth trajectories. These costs would be particularly challenging to nations that have scant resources compared to industrialized countries, and consider alleviating poverty as their first priority.

The Green Climate Fund (GCF) is a proposal by Parties to the United Nations Framework Convention on Climate Change (UNFCCC)[3] to establish an international financial institution to assist developing countries in their efforts to combat climate change. The fund would be capitalized by contributions from donor countries and other sources and used to support climate change mitigation and adaptation[4] projects, programs, policies, and other activities. The GCF would complement, or perhaps replace, many of the existing multilateral climate change funds (e.g., the Global Environment Facility, the Climate Investment Funds, the Adaptation Fund), and become the official financial mechanism of the UNFCCC. Expectations by many countries, specifically developing countries, are that the fund may be very large (i.e., in the range of several tens of billion to over a hundred billion dollars annually), and serve as the predominant institution for climate change assistance in the developing world.[5] These countries believe that the agreement to establish the GCF has been a key success in the recent international negotiations. But others caution that ambitious steps need to be taken to ensure that the fund is designed and implemented correctly in order to achieve an adequate buy-in by donor countries of its effectiveness and by recipient countries of its legitimacy.

The GCF is in the planning stage. Its scope, governance structure, sourcing, and disbursement mechanisms are currently under negotiation by Parties to the UNFCCC. Having been mentioned first in the UNFCCC-adopted Copenhagen Accord of December 2009, a decision was made to establish the fund during the UNFCCC 16th Conference of Parties (COP) in Cancun, Mexico, in December 2010. Negotiations have since turned to the design of the fund, over the period of March to November 2011, and under the auspices of a COP-appointed Transitional Committee, with the intent of bringing a finished proposal before the UNFCCC 17th COP in Durban, South Africa, November 28–December 9, 2011.

Many issues remain to be clarified during negotiations, and some involve long-standing and contentious debate. They include:

- what role the CGF would play in providing sustained finance at adequate levels,
- how it would fit into the existing development assistance and climate financing architecture,
- how it would be legally and institutionally governed,
- how it would be capitalized, and
- how it would allocate and deliver assistance efficiently and effectively to developing countries.

The United States is a Party to the UNFCCC negotiations. It also is represented on the UNFCCC COP-appointed Transitional Committee to negotiate the design of the CGF. The U.S. representative is Marisa Lago, Assistant Treasury Secretary for International Development and Markets, U.S. Department of the Treasury. Ms. Lago is supported by the U.S. Department of the Treasury's Office of Energy and Environment. If and when the Transitional Committee reaches consensus on a design for the fund, and if and when the proposal is brought before the COP for a decision, the U.S. Congress—through its role in authorizations, appropriations, and oversight—would have significant input on U.S. participation in the fund. Congressional input on a potential UNFCCC-established fund may include:

- whether and when to participate in the fund,
- how much to contribute to the fund, and with what source or sources of finance,
- whether fund contributions would carry specific guidance in distribution and use,
- how contributions to the fund would relate to other U.S. bilateral, multilateral, and private sector climate change assistance, and
- whether and when to consent to negotiated treaty obligations, if submitted.

After providing a brief background on the history of the climate negotiations leading up to the proposal of the GCF, this report (1) reviews the terms of the agreement for the design of the fund, (2) outlines outstanding design issues, and (3) addresses the relationship of the fund to other U.S. commitments for international climate change assistance. The report ends with a summary of congressional issues of relevance to the fund.

Background on the United Nations Framework Convention on Climate Change

The UNFCCC was the first formal international agreement to acknowledge and address human-driven climate change. The U.S. Senate provided its advice and consent to the Convention's ratification in 1992, the same year it was concluded.[6] For the United States, the UNFCCC entered into force in 1994. As of January 2011, 194 governments are Parties. As a framework convention, the UNFCCC provides a structure for international consideration of climate change but does not contain detailed obligations for achieving particular climate-related objectives in each Party's territory. It recognizes that climate change is a "common concern to humankind," and, accordingly, requires parties to (1) gather and share information on GHG emissions, national policies, and best practices; (2) launch national strategies for addressing GHG emissions and adapting to expected impacts; and (3) cooperate in preparing for the impacts of climate change. The UNFCCC did not set binding targets for GHG emissions; however, it did commit the higher-income Parties (i.e., those listed in Annex II of the Convention)[7] to provide unspecified amounts of financial assistance to help lower-income countries meet the broad, qualitative obligations common to all Parties.[8]

As the treaty entered into force and the UNFCCC Conference of the Parties (COP) met for the first time in 1995, the Parties agreed that achieving the objective of the UNFCCC would require new and stronger GHG commitments. As a first step toward meeting this objective, the 1997 Kyoto Protocol was drafted and entered into force with a stated aim to reduce the net GHG emissions of industrialized country Parties (Annex I Parties) to 5.2% below 1990 levels in the period of 2008 to 2012. The United States signed the Kyoto Protocol in December 1997.

However, at the time, opposition in Congress was strong.[9] The Kyoto Protocol was not submitted to the Senate by President Clinton, or by his successor, President George W. Bush.

In 2007, UNFCCC Parties reconvened negotiations for further commitments beyond the Kyoto Protocol, and agreed to negotiate a suite of agreements that included new GHG mitigation targets for Annex I Parties, "nationally appropriate mitigation actions" for non-Annex I Parties, and other commitments for the post-2012 period. The mandates (referred to as the Bali Action Plan) specified that the products of negotiation should be ready by the end of 2009. Due perhaps to high expectations, as well as continued divergence between Parties on some key issues, the 2009 COP in Copenhagen, Denmark, did not produce a legally binding treaty, but a short, non-legally binding political document called the Copenhagen Accord.[10]

The Copenhagen Accord was a policy document drafted among leaders of about two dozen countries in the final hours of the 2009 COP, and subsequently acknowledged by 114 countries. The Accord sat in sharp contrast to the Kyoto Protocol negotiations, as its bottom-up and nationally appropriate model differed greatly from the top-down implementation of the Protocol. Provisions in the Accord included voluntary GHG mitigation efforts by all Parties, adaptation and forestry actions, technology transfer mechanisms, and transparency and reporting standards, as well as a joint pledge by developed countries of up to $30 billion in "fast start" climate financing for developing countries for the period 2010–2012, and a goal of mobilizing $100 billion annually by 2020. The Accord also established the "Green Climate Fund" (GCF) to serve as the operating entity of the mechanism of the Convention.

Many of the elements of the Copenhagen Accord, the Bali Action Plan, and the UNFCCC were adopted officially at the 2010 COP in Cancun, which yielded several decisions collectively called the Cancun Agreements.[11] Climate financing was one aspect of the negotiations.

The Cancun Agreements on Climate Finance

The establishment of the Green Climate Fund—as well as some other financial arrangements mentioned in the Copenhagen Accord—was decided on by Parties at the 2010 COP in Cancun, Mexico, and entered into the negotiating text of the UNFCCC's Ad Hoc Working Group on Long-term Cooperative Action under the Convention (AWG-LCA).[12] Selected provisions include:

Financial Pledges

- Fast Start Financing. The agreement puts forth a collective commitment by developed country Parties (not specified in the text) to provide new and additional resources approaching $30 billion for the period 2010–2012 to address the needs of developing countries (the allocation, or "burden-sharing," among countries was not specified in the text) (§ 95).
- 2020 Pledge. The agreement takes note of the pledge by developed country Parties (not specified in the text) to achieve a goal of mobilizing jointly $100 billion per year by 2020 to address the needs of developing countries (the allocation, or "burden-sharing," among countries was not specified in the text) (§ 98).
- Sources. The agreement outlines that the pledged assistance is to be scaled-up, new and additional, predicable and adequate, and that it may come from a wide variety of sources, including public and private, bilateral, multilateral, and alternative (§§ 97, 99).[13]
- Balanced Package. The agreement recognizes that the financial pledges are offered in the context of continued negotiations toward a balanced package of commitments by all Parties that includes, among other items, meaningful actions on mitigation[14] and transparency[15] (§ 98).

The Green Climate Fund

- Green Climate Fund. The agreement opens the way for the establishment of the GCF, to be designated as an operating entity of the financial mechanism of the UNFCCC, accountable to and under the guidance of the COP, to support projects, programs, policies, and other activities in developing country Parties (§ 102).
- Thematic Funding Windows. The agreement states that the fund should employ thematic funding windows (i.e., sub-accounts targeted at specific climate change activities, such as mitigation, adaptation, forestry, capacity-building, etc.). The allocation among these categories was not specified in the text, only that it should be "balanced" between adaptation and mitigation (§ 102).

- Adaptation. The agreement notes that a "significant share" of any new multilateral funding for adaptation should flow through the GCF (§ 100).
- Governance. The agreement sets up a 24-member Board for the fund, comprising an equal number of members from developing and developed country Parties (not specified in the text), wherein representation from developing country Parties is specifically to include relevant U.N. groupings, including regions, small island States, and the least developed countries (§ 103).
- Trustee. The agreement designates that the fund has a trustee, separate from the Board and the UNFCCC, to manage the financial assets of the fund, maintain appropriate financial records, and prepare financial statements and other reports required by the Board, in accordance with internationally accepted fiduciary standards. The agreement invites the World Bank to serve as an interim trustee, subject to a review after three years of operation (§§ 104–107).
- Secretariat. The agreement calls for the creation of an independent Secretariat to support the operation of the fund, with guidance from the Board (§ 108).
- Transitional Committee. The agreement then stipulates the formation of a Transitional Committee to design the fund, comprising 40 members, with 15 members from developed country Parties and 25 members from developing country Parties, with experience and skills in the areas of finance and climate change, in accordance with given Terms of Reference (§§ 109–110).
- Technical Support Unit. The agreement then tasks the COP to make arrangements for supplying a support staff for the Transitional Committee, to be sourced from relevant U.N. agencies, international financial institutions, multilateral development banks, and the Global Environment Facility (§ 111).
- Standing Committee. Finally, the agreement establishes a permanent Standing Committee under the UNFCCC to assist the COP in exercising its functions with respect to the financial mechanism of the Convention. The roles and functions of the Standing Committee are yet to be defined by Parties, but the provision allows for further discussion on the legal relationship between the fund and the UNFCCC (§ 112).

The Transitional Committee

A Transitional Committee is tasked with developing an operational design for the GCF and recommending it for approval at the UNFCCC 17th COP in Durban, South Africa, from November 28 to December 9, 2011. The issues to be addressed by the Transitional Committee are outlined in a Terms of Reference section in the Cancun Agreements, and include:[16]

- legal and institutional arrangements for the fund,
- rules of procedure for the fund's Board, and the role of the Secretariat,
- financial instruments, funding windows, and access modalities to be employed by the fund,
- complementarity with other funds and institutions, and
- provisions for standards, safeguards, independent performance evaluations, technical advice mechanisms, and stakeholder participation.

DESIGN CHALLENGES OF THE FUND

Observers[17] have noted that the Green Climate Fund confronts many challenges in design, scope, governance, and implementation that must be negotiated by the Transitional Committee and, eventually, the UNFCCC Conference of Parties. Several of the most significant challenges include:

Timeline of the Negotiations

Many developing (i.e., recipient) country Parties and their civil society organizations have high expectations for the GCF. They are pushing for the Transitional Committee to deliver on several of the more long-standing and contentious points of negotiation (e.g., sourcing of funds, assessment of contributions, determination of burden-sharing, targets for disbursement). However, other Parties—including the United States—are insisting that the Transitional Committee concentrates solely on the technical design characteristics of the fund (as it was tasked to do by the Cancun Agreements) in order to deliver a timely and manageable product to the COP. Notwithstanding, the Terms of Reference for the CGF are extensive. The Transitional Committee met once on April 28 and 29, 2011, in Mexico City,[18] and has agreed to meet on three more occasions prior to the UNFCCC 17th COP in Durban, South Africa, from November 28 to December 9, 2011. Even in the best of circumstances, many believe that it would be difficult to successfully negotiate a design for the fund during this time frame, much less address the more long-standing and contentious issues. Some observers have suggested that it may be several years before the Transitional Committee can report a final design to the COP.

Scope of the Fund

While little has been decided regarding the eventual size and scope of the GCF, its formation is being viewed by many as a means through which to simplify the complex network of multilateral and bilateral funding mechanisms that currently provide climate change assistance to developing countries. Many early proponents of a global fund had envisioned that such an institution would play the role of a "fund of funds," or an "umbrella," under which to collect both the resources and the comparative advantages of the other mechanisms. As it currently stands, the negotiating text from the Cancun Agreements gives little indication that such an ambition is to be pursued by the GCF. It refers, instead, to "aligning" the GCF with other institutions. Nevertheless, the fate of existing multilateral financial mechanisms may be called into question by the establishment of the GCF. At present, the Adaptation Fund is the sanctioned U.N. mechanism in support of climate change assistance for adaptation actions. The Global Environment Facility is the sanctioned UNFCCC financial mechanism in support of mitigation actions. The World Bank's Climate Investment Funds were initially designed to sunset in 2012 at the presumed commencement of the new UNFCCC mechanism. It is possible that the eventual scope of the GCF may overshadow and/or replace these funds.[19] Conversely, it is also possible that the CGF, if established, may prove inadequate to existing arrangements in the eyes of potential donors.

Fund Governance

Many decisions on governance await the deliberations of the Transitional Committee. While the composition and selection of the Board and the Secretariat are important considerations, two other bodies have provoked significant debate: the Trustee and the Standing Committee.[20]

- The Trustee. The role of the World Bank in the GCF has been, and continues to be, controversial. The Cancun negotiating text invites the World Bank to serve as the interim trustee, subject to review after three years of fund operation (§ 107). Most believe that once established, a subsequent shift in institutional arrangement is doubtful. Many developing countries hold the World Bank in a negative light, believing it to be non-transparent, overly bureaucratic, and reflecting solely the interests of higher-income countries, which command greater decision-making power by virtue of their greater financial contributions. Additionally, some Parties see the potential for conflicts of interest during the design phase of the fund, since the Bank (1) already operates a portfolio of Climate Investment Funds that might compete against the GCF for potential donor country contributions, and (2) has been asked to serve as support staff for the Transitional Committee to aid in designing the operational procedures, project selection criteria, performance standards, and safeguard measures for the new fund. Despite these concerns, some Parties remain unconvinced that an adequate substitute exists, claiming that no other extant institution could undertake the proposed financial administration and fiduciary standards with the same level of confidence from the donors.
- The Standing Committee. The Cancun negotiating text establishes a committee (its size, composition, and participation are not specified in the text) to help the COP in exercising its functions with respect to the financing obligations in the Convention. The committee is specifically tasked to improve "coherence and coordination in the delivery of climate change financing, mobilization of financial resources and measurement, [and] reporting and verification of support provided ..." (§ 112). While currently vague, the roles and functions of the Standing Committee are to be defined further by Parties during AWG-LCA negotiations. Some[21] see the text on the Standing Committee to be a "place-holder" for postponing the more controversial issues until later. Others believe it was included to provide a direct, sustained, and authoritative link between the UNFCCC and the GCF Board, perhaps to open a channel for future mandates to flow directly from the COP to the Board.

Fund Mobilization

The Cancun negotiating text is silent on sources, with no proposal for how finance would flow into the fund. The text simply "takes note" of "relevant reports on the financing needs and options for mobilization of resources to address the needs of developing country Parties with regard to climate change adaptation and mitigation, including the report of the High-level Advisory Group on Climate Change Finance [AGF]" (§ 101). Most see the absence of this item under the Transitional Committee's Terms of Reference as unsurprising. The 2010 report by the AGF concluded that the goal of mobilizing adequate and predictable climate finance to developing countries on the order of $100 billion annually would be "challenging but feasible."[22] Further, while it is acknowledged that adequate international finance would likely require a range of sources (including public finance, development bank instruments, carbon markets, and private capital), little unity exists among COP Parties as to the balance between public and private sources, developed and developing country participation in international carbon markets or tax schemes, and the political feasibility of other large-scale fund mobilizations. Several other multilateral fora have taken up the issue of climate finance sourcing, including the G-20 and the Major Economies Meeting.[23] However, the prospect of the issue being resolved during the course of the Transitional Committee's deliberations appears to be slight.

Fund Disbursement

The Transitional Committee's Terms of Reference mention three main design aspects with respect to the disbursement of funds: "financial instruments, funding windows and access modalities" (Appendix III, § 1c). Each category engenders debate among Parties.

- **Instruments.** As for the choice of instruments, observers[24] stress that climate finance can take a variety of forms: it can be provided as grants, concessional loans, market-based loans, equity investments, or guarantees. However, debates consistently arise between donor and recipient countries as to the appropriateness of debt-based instruments (e.g., loans) for humanitarian aid. Whilst the general presumption is that climate finance in support of adaptation actions in developing countries should be provided on grant terms, this is less customary with regard to mitigation actions. Thus, many see it as important for the GCF to secure a good match between the type of finance and the object of financing, retaining sufficient funds to provide grants when necessary, as influenced by both the country and the project profile.

- **Windows**. While the Copenhagen Accord specified that the GCF would support activities related to "mitigation including REDD-plus,[25] adaptation, capacity building, technology development and transfer" (§ 10), the Cancun negotiating text has dropped such references, opting instead to state that the GCF would use "thematic funding windows" (§ 102). Whether or not the question of "balance" among these windows is addressed by the Transitional Committee remains to be seen. With present funding by existing financial institutions decidedly tilted toward supporting mitigation actions, there is likely to be a strong expectation—by developing countries as well as certain civil society organizations—that adaptation actions would receive a significant portion of support from the CGF. Some allocation formula is likely to be expected from the Transitional Committee, if for no other reason than to allow subsequent funding decisions to be adequately rule-based.
- **Access.** Finally, consideration of how countries would access funds from the GCF, and which agencies and organizations would be allowed to acquire funds to implement projects, remains an ongoing issue of debate. Currently, almost all multilateral financial assistance for climate change activities in developing countries is channeled through third-party implementing agencies (e.g., U.N. agencies, multilateral development banks, major nongovernmental organizations).[26] Whether and which of these agencies would be accredited to manage GCF funds is not specified; however, language in the Cancun negotiating text does mention one specific modality: "direct access" (Appendix III, § 1c). Direct access has become a prominent, new arrangement in climate finance delivery, allowing the recipient country to access financial resources directly from the fund, and/or allowing it to assign an implementing agency of its own choosing. This operational freedom has been a rallying point for many developing country Parties. The modality is also supported by many developed country Parties as a means to secure broader competition and greater country ownership.

- Nevertheless, implementation of direct access arrangements may prove to be slow and difficult, because they would likely require the same stringent level of fiduciary standards, competitive procurement practices, and environmental and social safeguards demanded of existing third-party implementing agencies.

Relationship of the Fund to Ongoing Negotiations

Many Parties have stressed the importance of the design choices to be made by the Transitional Committee, and the influence that these choices may have on the ability of the fund to attract both the longer-term financial pledges of donor countries and the leveraged investment of the private sector. A few, including the United States, contend that the engagement of finance ministries—as opposed to climate negotiators—is critical to ensuring the best technical design and the best use of funds. Some observers[27] have cautioned that a poorly designed fund could cause the multilateral effort to deteriorate into endless political squabbles as opposed to effective technical conversations about how best to capitalize and leverage resources. Nevertheless, while some critics see climate negotiators as ill-equipped to adequately debate issues of finance, their expertise remains relevant to other aspects of the negotiations. Many Parties remain focused on the need for a "balanced package" of commitments to arise from the UNFCCC climate negotiations. They, including the United States, stress that the design of a financial mechanism should not outpace elaborations on other issues, such as GHG mitigation and transparency commitments, to be made by developed and developing countries alike.

Relationship of the Fund to the Convention

As currently conceived, the GCF is intended to operate at arm's length from the UNFCCC, with an independent Board, Trustee, and Secretariat. The Cancun negotiating text states that the GCF is to be "accountable to and function under the guidance of the Conference of Parties" (§ 102) (i.e., similar in legal structure to the Global Environment Facility), as opposed to "accountable to and function under the guidance *and authority* of the Conference of Parties" (i.e., similar in legal structure to the Adaptation Fund). While subtle, the distinction carries import, and negotiators from China and the Group of 77[28] have—for the moment at least—kept the latter structure in conversation in an effort to ensure adequate representation by all Parties of the UNFCCC. The majority of developed country Parties, however, oppose the Adaptation Fund model as inefficient and overly politicized for two key reasons: (1) the COP would have direct authority over the selection and release of all Board and Secretariat members, and (2) the COP would have final approval over all rules and guidelines proposed by the Board. Given the current language of the negotiating text, and the proposed design of the Board to carry equal representation between developed and developing country Parties, this issue may already be resolved.

Relationship of the Fund to Other U.S. Climate Finance Commitments

At present, there have been no funds pledged specifically to the Green Climate Fund. Further, there have been no pronouncements regarding redirection of current or future bilateral or multilateral financial assistance into the fund. Nor has the size, the scope, or the sources of financial contributions been decided in the negotiating text or voiced officially by the UNFCCC, Parties to the COP, or members of the Transitional Committee. The only specific financial allocation tied directly to the GCF in the Cancun negotiating text is "that a significant share of new multilateral funding for adaptation *should* flow through the Green Climate Fund" (§ 100, italics mine). This statement does not define "new" or "significant," does not identify specific contributors or recipients, and does not assign specific amounts, allocations, or changes to bilateral aid, mitigation assistance, or current levels of funding. However, its inclusion may be the result of Parties' experiences with the U.N. Adaptation Fund, which serves as a reminder of the challenge to raise funds directly from countries' national budgetary contributions and the reluctance of such countries to commit a significant share of these funds to UNFCCC-sponsored adaptation projects.

In regard to the GCF's relationship to other climate finance commitments by the United States:

- **UNFCCC Fast Start Financing Pledges.** The collective pledge by developed country Parties to provide new and additional resources approaching $30 billion for the period 2010–2012 is *not* tied to the

GCF. In fact, the GCF is unlikely to be operational before the end of 2012. Fast start funds most likely would continue as current bilateral and multilateral contributions made through authorized appropriations by donor country governments.

- UNFCCC 2020 Pledges. The collective pledge by developed country Parties to the goal of mobilizing jointly $100 billion per year by 2020 is *not* tied directly to the GCF. The Cancun negotiating text makes clear that "funds provided to developing country Parties may come from a wide variety of sources, public and private, bilateral and multilateral, including alternative sources" (§ 99). If established, the GCF would be one of many possible public and multilateral sources. While any financial assistance that is channeled through the GCF would likely be considered a part of the $100 billion goal, the entirety of the $100 billion goal is not expected to be provided solely by the GCF, and no estimation of the GCF's presumed share has been suggested. Many Parties, as well as the AGF report, have suggested that development bank instruments, carbon markets, and—especially—private capital would be critical to mobilizing assistance at the level pledged.
- Bilateral Aid. The GCF would not necessarily interfere with current or proposed bilateral climate change assistance to developing countries. The GCF would be another multilateral mechanism for climate change assistance that would exist alongside bilateral activities, much the way that the Global Environment Facility and the Climate Investment Funds currently do. U.S. allocations between and among bilateral and multilateral assistance channels would continue through authorized congressional appropriations.
- Other Multilateral Aid. The GCF Transitional Committee and the GCF Standing Committee have been tasked with determining the complementarity of the GCF with respect to other U.N. multilateral mechanisms. Thus, the negotiations may produce some alteration in the landscape of the multilateral choices provided by the UNFCCC. Development bank mechanisms such as the Climate Investment Funds may or may not be reevaluated by their governing Boards in light of any proposal that comes out of the UNFCCC. Presumably, choices would remain available to donor countries. U.S. allocations among multilateral assistance channels would remain based on congressional guidance and would continue through authorized congressional appropriations.

ISSUES FOR CONGRESS

Members of Congress hold mixed views about the value of international financial assistance to address climate change. While some Members are convinced that human-induced climate change is a high-priority risk that must be addressed through federal actions and international cooperation, others are not as convinced. Some are wary, as well, of international processes that could impose costs on the United States, redirect funds from domestic budget priorities, undermine national sovereignty, or lead to competitive advantages for other countries. Regardless of current views, the United States is a Party to the UNFCCC and has certain obligations under the treaty. The executive branch continues negotiations and implementation of the UNFCCC obligations, while committees of Congress engage in oversight (from home and at the international meetings), providing input to the executive branch formally and informally, and deciding program authorities and appropriations for these activities.

As Congress considers potential authorization and/or appropriations for the Green Climate Fund, if established, it may raise concerns regarding the cost, purpose, direction, efficiency, and effectiveness of the UNFCCC and existing international financial institutions. These concerns may be weighed against the proposed design characteristics of the new fund in an effort to assess its potential performance. Congress may then be required to determine the allocation of funds between bilateral and multilateral climate change assistance as well as among the variety of multilateral mechanisms. Congress may also wish to gauge and give guidance to the new fund's relationship with domestic industries and private sector investment, as well as the spillover effects of U.S. participation on technological innovation, humanitarian efforts, national security, and international leadership. Potential authorizations and appropriations for the GCF rest with several committees, including the U.S. House of Representatives Committees on Foreign Affairs (various subcommittees); Financial Services (Subcommittee on International Monetary Policy and Trade); and Appropriations (Subcommittee on State, Foreign Operations, and Related Programs); and the U.S. Senate Committees on Foreign Relations (Subcommittee on International Development and Foreign Assistance, Economic Affairs, and International Environmental Protection); and Appropriations (Subcommittee on State, Foreign Operations, and Related Programs).

Additional issues for Congress concerning the climate negotiations in general, and the GCF in particular, may include the means to establish a more desirable form of agreement; the compatibility of any international agreement with U.S. domestic policies and laws; the adequacy of appropriations and fiscal incentives to achieve any commitments under the agreement; and any requirements for potential ratification and implementing legislation, should a formal treaty emerge from the negotiations.

End Notes

[1] This report assumes a general understanding of climate change science, policy, financing, and international negotiations. For further background on climate change science and policy, see CRS Report RL34266, *Climate Change: Science Highlights*, by Jane A. Leggett, and CRS Report RL34513, *Climate Change: Current Issues and Policy Tools*, by Jane A. Leggett; for further background on international climate change financing, see CRS Report R41808, *International Climate Change Financing: Needs, Sources, and Delivery Methods*, by Richard K. Lattanzio and Jane A. Leggett; and for further background on the international climate change negotiations, see CRS Report R40001, *A U.S.-Centric Chronology of the International Climate Change Negotiations*, by Jane A. Leggett.

[2] Most industrialized countries currently deliver some financial and technical assistance through a variety of bilateral development programs and multilateral financial institutions. "Bilateral" assistance involves direct transfers from one country to another; "multilateral" assistance is distributed through international organizations and agencies such as the United Nations and the World Bank Group.

[3] The UNFCCC and its processes are discussed in greater detail in the next section, "Background on the United Nations Framework Convention on Climate Change".

[4] "Mitigation activities" refer to actions taken to reduce or reverse the forces that contribute to global climate change (examples include transitioning to a low-emissions energy supply; capturing the opportunities in energy efficiency improvements in buildings, transportation, and industry; reducing deforestation and improving sustainable forest management to better serve as GHG emissions sinks; and employing more low-emissions and sustainable agriculture practices). "Adaptation activities" refer to adjustments made in natural or human systems in response to actual or expected climate change and its effects (examples include employing climate-resistant crop varieties, improving irrigation systems, integrating sustainable land management into agricultural planning, protecting water resources, managing coastal zones, designing infrastructure for extreme weather or for sea-level rise, and improving public health services).

[5] Many developing countries have stated their support for a large and centralized fund for climate change assistance to be housed at the UNFCCC, capitalized primarily by public contributions, and funded at or near the level of the estimated costs for climate change activities in the developing world. While cost estimates vary greatly depending upon the analysis (see "Cost Estimates" section in CRS Report R41808, *International Climate Change Financing: Needs, Sources, and Delivery Methods*, op cit.), many countries have acknowledged the annual $100 billion figure as a target. To put this figure in context, in FY2010 the United States provided $1.3 billion for international climate change assistance, split almost equally between bilateral and multilateral programs. Further, the FY2010 U.S. budget authority for *all* foreign operations programs, both bilateral and multilateral, at the Departments of State, Treasury, and the Agency for International Development, totaled

$32.8 billion, or approximately 3% of all FY2010 U.S. discretionary spending. Beyond the existing multilateral funds, total global contributions to international climate change assistance have been difficult to track due to the lack of international standards. Transparency of financial contributions and standardized accounting procedures are included in the package of agreements currently under negotiation by Parties to the UNFCCC.

[6] United Nations Framework Convention on Climate Change, Treaty Number: 102-38, October 7, 1992, the resolution of advice and consent to ratification agreed to in the Senate by Division Vote.

[7] UNFCCC Annex I Parties include the industrialized countries that were members of the OECD (Organization for Economic Cooperation and Development) in 1992, plus countries with economies in transition (the EIT Parties), including the Russian Federation, the Baltic States, and several Central and Eastern European States. Annex II Parties consist of the OECD members of Annex I, but not the EIT Parties.

[8] For more information on the UNFCCC and U.S. participation in international climate treaties, see CRS Report R41175, *International Agreements on Climate Change: Selected Legal Questions*, by Emily C. Barbour.

[9] See the Byrd-Hagel Resolution (S.Res. 98) in July 1997, wherein the Senate expressed its opposition (95-0 vote) to the terms of the Berlin Mandate (the 1995 UNFCCC COP agreement that led to the adoption of the Kyoto Protocol) by stating that the United States should not sign any treaty that does not include specific, scheduled commitments of non-Annex I Parties in the same compliance period as Annex I Parties, or that might seriously harm the U.S. economy.

[10] See UNFCCC Decision 2/CP.15, Copenhagen Accord, FCCC/CP/2009/11/Add.1, at http://un fccc.int/resource/docs/ 2009/cop15/eng/11a01.pdf.

[11] See UNFCCC Decision 1/CP.16, The Cancun Agreements: Outcome of the work of the Ad Hoc Working Group on Long-term Cooperative Action under the Convention, FCCC/CP/2010/7/Add.1 at http://unfccc.int/resource/docs/2010/ cop16/eng/07a01.pdf#page=2.

[12] The Ad Hoc Working Group on Long-term Cooperative Action under the Convention (AWG-LCA) is the negotiating track of the COP that serves "to enable the full, effective and sustained implementation of the Convention through long-term cooperative action, now, up to, and beyond 2012." This is in contrast to the Ad Hoc Working Group on Further Commitments for Annex I Parties under the Kyoto Protocol (AWG-KP) (for which the United States is not a Party, but an observer). Both Working Groups are tasked with producing negotiating texts with the aim of moving the texts to an agreement among Parties to the UNFCCC COP.

[13] The United Nations convened an advisory panel in 2010 to investigate issues regarding the sourcing of climate finance. The report by the U.N. High Level Advisory Group on Climate Change Financing was released in November 2010 and can be found at http://www. un.org/wcm/content/site/climatechange/pages/financeadvisorygroup. For a discussion of the various proposed public, private, and alternative sources, as well as the issues revolving around the mobilization of international climate finance, see also CRS Report R41808, *International Climate Change Financing: Needs, Sources, and Delivery Methods*, by Richard K. Lattanzio and Jane A. Leggett.

[14] "Mitigation" commitments refer to the formalized pledges taken by developed country Parties, or the nationally appropriate measures being taken by developing country Parties, to reduce GHG emissions.

[15] "Transparency" commitments refer to the negotiated provisions whereby all major economy Parties (including the large emerging economies) must report on the progress they are making in meeting their mitigation commitments (targets and actions), and all Parties providing financial assistance must report their contributions through commonly accepted formats.

[16] The Terms of Reference for the design of the Green Climate Fund can be found in Appendix III of the Cancun Agreements, op. cit.

[17] Many economic and environmental civil society organizations have watched and reported on the UNFCCC climate negotiations in general, and the GCF negotiations in particular. For more analyses on the design challenges outlined in this report, among others, as well as on other proposed fund models, see, for example, Neil Bird, Jessica Brown, and Liane Schalatek, *Design Challenges for the Green Climate Fund, Climate Finance Policy Brief No. 4*, Heinrich Böll Foundation North America and Overseas Development Institute, January 2011; Nigel Purvis and Andrew Stevenson, "Climate Negotiations and International Finance," *Resources*, Winter/Spring 2011, No. 177, pp. 16-18; and Benito Müller, *Time to Roll Up the Sleeves—Even Higher!: Longer-term Climate Finance after Cancun, Oxford Energy and Environment Brief*, Climate Strategies and the Oxford Institute for Energy Studies, January 2011.

[18] The April meeting of the Transitional Committee succeeded in electing three co-chairs to the committee (Mexico, South Africa, and Norway), as well as dividing the Terms of Reference into four tentative clusters of issues, or work-streams (Scope, guiding principles, and cross-cutting issues; Governance and institutional arrangements; Operational modalities; and Monitoring and evaluation), that were then delivered to relevant support staff in the Technical Support Unit, under the guidance of co-chairs, to be researched and reported on by the next Transitional Committee meeting in Tokyo, Japan, in July.

[19] Recent funding levels for the above-mentioned multilateral funds are as follows: Global Environment Facility, $3.5 billion pledged for the period 2011-2014; Climate Investment Funds, $6.1 billion pledged for the period 2009-2012; Adaptation Fund, based on a formula of 2% of the Certified Emission Reduction units issued for projects of the Clean Development Mechanism of the UNFCCC Kyoto Protocol.

[20] During the April 2011 Meeting of the Transitional Committee, it was reported that Nicaragua caused a stir by accusing the World Bank of "Arthur Andersen syndrome," presumably referencing the accounting firm's responsibility in the Enron affair. See Lisa Friedman, "Negotiations: Nations begin to plan details of Green Climate Fund," *ClimateWire*, May 2, 2011. Similarly, the language regarding the Standing Committee produced considerable debate among negotiators during the UNFCCC intersessional meeting in Bonn, Germany, from June 6-17, 2011.

[21] Bird, op cit., p. 4.

[22] See AGF Report, op cit.

[23] The "G-20" refers to the Group of Twenty: a forum for finance ministers and central bank governors established in 1999 to bring together systemically important industrialized and developing economies to discuss key issues in the global economy. The "Major Economies Forum" refers to a forum of 17 major developed and developing economies established in 2009 to facilitate dialogue on energy and climate issues.

[24] Bird, op cit., p. 6.

[25] "REDD-plus," or "Reducing Emissions from Deforestation and Forest Degradation," activities refer to mitigation-relevant activities in support of forestry and sustainable land management.

[26] The only current exception to this practice is the U.N. Adaptation Fund, which has begun to accredit national agencies in recipient countries as official implementing agencies for fund disbursement. Council Members for the Global Environment Facility have begun discussions on a similar practice. This practice is referred to as "direct access."

[27] For example, see comments made by Athena Ballesteros of World Resources Institute in an article by Lisa Friedman, "Cancun: Countries Get to Work Designing Green Climate Fund," *ClimateWire*, January 3, 2011.

[28] The Group of 77 is an official U.N. negotiating group composed of developing countries. Founded in 1964 in the context of the U.N. Conference on Trade and Development (UNCTAD), it now has over 130 members.

CHAPTER SOURCES

The following chapters have been previously published:

Chapter 1– is an edited, reformatted and augmented version of Congressional Research Service Report R41808, dated May 5, 2011.
Chapter 2– is an edited, reformatted and augmented version of Congressional Research Service Report R41845, dated June 1, 2011.
Chapter 3– is an edited, reformatted and augmented version of Congressional Research Service Report R41302, dated May 5, 2011.
Chapter 4– is an edited, reformatted and augmented version of Congressional Research Service Report R41889, dated June 23, 2011.

INDEX

F

G

H

I

J

K

L

M

N

O

P

Q

R

S

T

U

V

W

Y